IRON BACTERIA

IRON BACTERIA

BY

DAVID ELLIS

D.Sc., Ph.D., F.R.S.E.

ROYAL TECHNICAL COLLEGE, GLASGOW

WITH 45 ILLUSTRATIONS AND 5 PLATES

NEW YORK
FREDERICK A. STOKES COMPANY
PUBLISHERS

PREFACE

IT is nowadays increasingly felt that there should
be greater co-ordination between the work of the
biologist and the application of this work by engineers and chemists. This work has two ends in view.
The first is to bring forward for consideration the present state of knowledge, and bring within the compass
of a small book the main contributions from this and
other countries. This, it is hoped, will facilitate the
work of future investigators and indicate where the
gaps in our knowledge are greatest. The present
writer has been engaged on the investigation of the
iron-bacteria for a period of ten years and is well
aware of the many points which still need patient
research for their elucidation. While most of the
material in the book has appeared at different times
in various journals, a good part appears now for the
first time. It is hoped that the present volume will
attract the notice of biological investigators, and induce one here and there to take up the study of the
interesting group of organisms included under the
iron-bacteria. To such the book will serve as a

point of concentration from which to launch further attacks. Most biologists would probably confess to a very scanty knowledge of these organisms, partly because they are not adequately treated in most text-books on biology, and partly because for efficient knowledge of them a special training in bacteriology must be superposed on a training along general biological lines. To any such, therefore, who desire a more extended knowledge of this particular class of organisms than is possible in a general text-book, this book should prove of some help.

The second end in view is to reach the notice of the water engineer and the analytical chemist, and perhaps to elicit a measure of sympathy for the endeavours of the biologist. The work of the writer brings him constantly into contact with practical problems of applied biology. There is still a lingering doubt in the mind of the " practical man " that science and practice are antagonistic, and in the work of the past, in so far as this touches the iron-bacteria, the outcome has not been the best that was possible, because problems have been attacked divorced from biological considerations. This has been especially the case in this country, and in the past we have lagged behind other countries in our appreciation of the work of the laboratory. It seems clear, for example, that certain effects due to the activities of organisms like the iron-bacteria cannot be adequately studied apart from the study of the organisms themselves.

The practical aim of the book is, therefore, chiefly to give indications to the engineer and to the chemist of the kind of organism that they are dealing with in certain cases, and to point out the principles which should underlie the method of treatment to remove the particular evil which has attacked their reservoir or their conduit pipes. Hitherto the sources of information at the disposal of our engineers and chemists have been very scanty, and these are chiefly scattered in various journals, of which very few are written in English. The photomicrographs are sufficiently clear to ensure identification of any of the iron-bacteria if a high-powered microscope be employed. An endeavour has been made to photograph the organisms in the most typical conditions, and when most suitable for purposes of identification.

The author is indebted to London University for a Grant from the Dixon Fund to safeguard him, partially at any rate, from any pecuniary loss incurred by the publication of this work.

DAVID ELLIS

ROYAL TECHNICAL COLLEGE
GLASGOW

TABLE OF CONTENTS

ix

CHAPTER IX

LIST OF PLATES

LITERATURE

ADLER, O. (1) Ueber Eisenbakterien in ihrer Beziehung zu den therapeutisch verwendeten natürlichen Eisenwässern. Centralblatt für Bakteriologie. Abth. II, Bd. II. 1903.

BEYTHIEN, HEMPEL, u. KRAFT. (1) Beiträge z. Kenntnis des Vorkommens von Crenothrix polyspora in Brunnenwässern. Zeitschr. f. Unters. d. Nahrungs. u. Genussmittel. 1904.

BINAGHI. (1) Instituto d'igiene della R. Università di Cagliari. 1913. Tubazioni in Ghisa ed altri materiali contenenti Ferro studiati in rapporto alla formazione dei tubercoli ferruginosi e al rammollimento del materiale delle condutture.

BROWN, CAMPBELL. (1) Deposits in Pipes and other Channels conveying Potable Water. Proc. of Institution of Civil Engineers. 1903-04. Part II.

BÜSGEN, M. (1) Kulturversuche mit. Cladothrix dichotoma. Berichte der deutschen Bot. Ges. Band 12. 1894.

CASAGRANDI, O. (1) Instituto d'igiene della R. Università di Cagliari. 1913.

COHN, F. (1) Ueber den Brunnenfäden Beiträge zur Biol. der Pflanzen. Bd. I, Heft 1. 1870.

EHRENBERG, D. C. G. (1) Die Infusionstierchen als vollkommene Organismen. Leipzig, 1838.

ELLIS, D. (1) On the discovery of a new genus of Thread-Bacteria (Spirophyllum ferrugineum, Ellis). Proc. Roy. Soc., Edinburgh. Vol. 27. 1907.
  (2) A contribution to our knowledge of the Thread-Bacteria (I). Centralblatt für Bakteriologie. Bd. 27. 1907.
  (3) Iron-Bacteria and their connection with Stone-decay. Proc. Roy. Philos. Soc., Glasgow. 1907.
  (4) A preliminary notice of five new species of Iron-Bacteria. Proc. Roy. Soc., Edinburgh. Vol. 28, Part V. 1908.
  (5) A contribution to our knowledge of the Thread-Bacteria (II). Centralblatt für Bakteriologie. Abth. II, Bd. 26. 1910.
  (6) On the new genus of Iron-Bacteria (Spirophyllum ferrugineum). Proc. Roy. Soc., Edinburgh. Vol. 31. 1911.

(7) An investigation into the life history of Cladothrix dichotoma. Proc. Roy. Soc., London. 1912.

(8) On the identity of Leptothrix Meyeri (Ellis) and Megalothrix discophora (Schwers) with Crenothrix polyspora. Centralblatt für Bakteriologie. Abth. II, Bd. 38. 1913.

(9) Fossil Micro-organisms from the Jurassic and Cretaceous Rocks of Great Britain. Proc. Roy. Soc., Edinburgh. Vol. 35. 1915.

FISCHER, A. (1) Untersuchungen über Bakterien. Pringsheim's Jahrb. für Wissench. Botanik. Bd. 27, Heft 1. 1895.

FLÜGGE. (1) Mikroorganismen. Aufl. 3. Leipzig, 1896.

GAIDUKOV, N. (1) Ueber die Eisen alga Conferva und die Eisenorganismen des Süsswassers in allgemeinen. Berichte d. Deutsch Bot. Ges. 1905.

GARRETT, J. H. (1) Public Health. 1896. Crenothrix Polyspora var. Cheltonensis.

GIARD, A. (1) Sur le Crenothrix Kühniana (Rabenhorst) cause de l'infection des eaux de Lille. Comptes rendus, etc. 1882.

HANSGIRG, A. (1) Prodromus der Algenflora von Böhmen. Theil 2. Archiv. für naturw. Landes-durch-forschung von Böhmen. Theil 8. Nr. 4. Prag., 1892.

HERDSMAN, W. H. (1) On the Organic Origin of the Sedimentary Ores of Iron and of their Metamorphosed Forms: the Phosphoric Magnetites. Journal of the Iron and Steel Institute. 1911.

HOEFLICH, K. (1) Kultur u. Entwicklungsgeschichte der Cladothrix Dichotoma (Cohn). Österr. Monats-schrift f. Tierheilkunde, etc. XXV. 1901.

HUEPPE. Naturw. Einführung in die Bakteriologie. Wiesbaden, 1896.

JACKSON, D. D. (1) A New Species of Crenothrix (Crenothrix manganifera). Trans. of the American Micros. Soc. Vol. XXIII. 1902.

(2) Hygienische Rundschau. 1904.

KENDALL, J. D. (1) The Iron-Ores of Great Britain. London, 1893.

KÜTZING. Phytologia generalis. 1843.

METTENHEIMER, C. (1) Ueber Leptothrix ochracea u. ihre, Beziehungen zu Gallionella ferruginea. Abhandl. der Senckenberg Naturforsch. Gesells. Bd. II. Frankfurt, 1856-58.

MIGULA, W. (1) Chlamydobacteriaceæ in Engler and Prantl's Die Pflanzenfamilien. 1896.

(2) Ueber Gallionella ferruginea. Berichte der deutsch. Bot. Ges. Bd. XV. 1897.

(3) System der Bakterien. Bd. II. Jena, 1899.

MOLISCH, H. (1) Die Pflanze in ihrer Beziehungen zum Eisen. Jena, 1892.

(2) Die Eisen-Bakterien. Jena, 1910.

MUMFORD, E. M. (1) A New Iron-bacterium. Journal of the Chemica Society. Vol. CIII. 1913.

RABENHORST. (1) Kryptogamen Flora von Schlesien. 1878.

(2) In Hedwigia. 1854.

RAUMER, E. (1) Ueber das Auftreten von Eisen und Mangan in Wasserleitungswasser. Zeitscr. f. analytische Chemie. 1903.

RÖSSLER, O. (1) Ueber Kultivierung von Crenothrix polyspora aus festem Nährboden. Archiv der Pharmazie. Bd. 233. Berlin, 1895.

(2) Der Nachweis von Crenothrix polyspora im Trinkwasser. Deutsche medizinische Wochenschrift. 1906. Nr. 40.

RULLMANN, W. (1) Die Eisen-bakterien in Lafar's techn. Mykologie. Bd. III.

(2) Ueber Eisen-bakterien. Centralblatt für Bakteriologie. Abth. II, Bd. 33. 1912.

SAVAGEAU ET RADAIS. (1) Sur les genres Cladothrix, Actinomyces et description de deux Streptothrix nouveau. Comptes rend. de l'acad., etc. 1892. Bd. 114.

SCHORLER, B. (1) Beiträge der Kenntnis der Eisenbakterien. Centralbl. für Bakt. Abth. II, Bd. XII. 1904.

SCHWERS, H. (1) Le fer dans les eaux Souterraines. Paris, 1908.

(2) Megalothrix discospora, eine neue Eisenbakterie. Centralblatt für Bakteriologie. Abth. II, Bd. 33.

SORLEY, H. C. (1) Proceedings of the Geological and Polytechnic Society, West Riding, Yorkshire, 1856-57.

WINOGRADSKY, S. (1) Ueber Eisenbakterien Bot. Zeitung. 1888.

ZOPF. (1) Untersuchungen über Crenothrix polyspora. Berlin, 1879.

(2) Zur Morphologie der Spaltpflanzen. Leipzig, 1882.

(3) Die Spaltpilze. Auflage III. 1884.

INTRODUCTION

T H E capacity for abstracting iron from the waters
in which they live and collecting it in the form
of ferric hydroxide on their surfaces is possessed
by various representatives of most of the classes of
micro-organisms that inhabit fresh waters. This capacity is a feature of certain protozoa, of certain algæ,
and of certain thread-fungi. It is also characteristic of
certain representatives of the large class of organisms
denominated under the term bacteria. It is to such
that the term iron-bacteria is applied. In the case
of these organisms, the phenomenon is more marked
because they not only collect more assiduously but
also have excellent structures for storing the ultimate
product, namely, ferric hydroxide, in the mucilaginous
sheaths which surround their bodies. It is to the
ferric hydroxide stored on and in the dead membranes
of iron-bacteria that in most cases is due the familiar
ochre-beds of ferruginous pools and streams. The
deposition on each membrane is often so great that
it exceeds the volume of the organism itself. Further,
the membrane is preserved from the dissolution which

IRON BACTERIA

CHAPTER I

LEPTOTHRIX OCHRACEA (*Kützing*)

Syn. CHLAMYDOTHRIX OCHRACEA (*Migula*)

THIS is by far the most widely distributed of all the iron-bacteria. The ochre-coloured deposit forming the beds of ferruginous streams which are so common in some districts in this country, is in almost all cases caused by the deposition of ferric hydroxide on the dead bodies of Leptothrix ochracea. This organism has been found in ferruginous streams in every country in which search has been made for it. Molisch (2) records that in many moorland meadows in *Austria* and in *Hungary* the water appears to be full of an ochre-coloured mass which is made up entirely of Leptothrix threads. The same author records that the colour of the curative ferruginous waters in Kitzbükel (Tyrol) is due entirely to a deposit of ferric hydroxide on countless threads of Leptothrix ochracea. The water of the river Moldau in its course through S.W. Bohemia before it joins the Elbe, flows through long stretches of moorland country and assumes a brownish colour in consequence of the large amount of organic matter which it takes up in solution.   In this brownish water, both in the Moldau and the Elbe, Molisch has found active growths of Leptothrix ochracea in company with other iron-bacteria.   Investigators in *Belgium* record the same wide distribution in that country.   Thus Schwers (1) has examined the deposits of 208 iron waters and has found Leptothrix to be the dominant organism in 142 of the samples.   In Great Britain also Leptothrix ochracea has an equally wide distribution.   In the West of Scotland, and particularly in the S.W. portion, ferruginous streams with ochre-coloured deposits are extremely common, and in practically all of them Leptothrix ochracea is the dominant organism.   I have examined samples not only from various parts of Great Britain, but also from Russia, Spain, and Denmark.   Altogether some 300 samples of the deposits of ferruginous streams were examined.   In 90 per cent of them Leptothrix ochracea was present, very often to the practical exclusion of all other organisms.   All the samples were drawn from the deposits of running or stagnant waters of a ferruginous nature.   Finally Molisch records that iron-waters were examined by him in Ceylon, Java, China, Japan, and America (Chicago), and that the contents of these were in nowise different from those of the European ferruginous streams.

This organism is therefore universally distributed in ferruginous waters in every country throughout the

PLATE I

FIG. 1. × 350

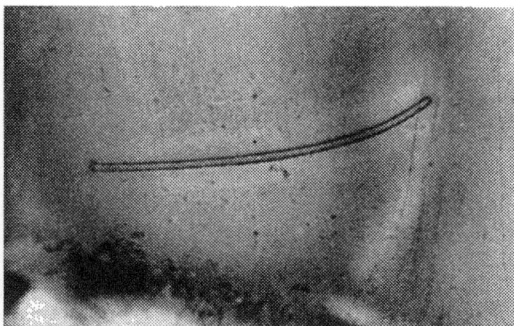

FIG. 2. × 350

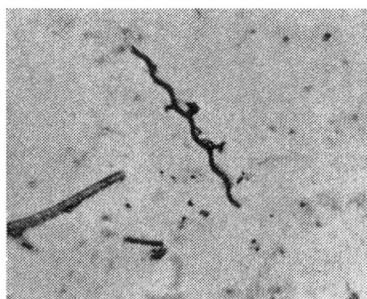

FIG. 3. × 350

LEPTOTHRIX OCHRACEA

FIG. 1. APPEARANCE TYPICAL OF OCHRE-BED DEPOSIT WHEN VIEWED THROUGH MICROSCOPE. THE STRAIGHT OR SOMEWHAT BENT THREADS ARE LEPTOTHRIX. IN THE FIELD ARE ALSO A FEW BANDS—THOSE SHOWING SPIRAL TWISTING—OF SPIROPHYLLUM. ALL THE ORGANISMS ARE DEAD, AND THE STRUCTURES ARE THEIR IRON-ENCRUSTED REMAINS

FIG. 2. SPECIMEN SHOWING SLIGHT CURVATURE

FIG. 3. SPECIMEN SHOWING THE BODY THROWN INTO A NUMBER OF REGULAR WAVE LENGTHS

world. No records appear of its presence in non-ferruginous waters, but we must not conclude that it is necessarily confined to iron-containing waters, for, as will be shown later, Leptothrix can be cultivated artificially in a medium totally devoid of iron. Hitherto, however, in nature, it has not been found in waters devoid of iron in solution.

Nomenclature.—This organism appears under three names :—

1. *Leptothrix ochracea.* — Bestowed on it by Kützing (1) in 1843. Leptothrix was an algal genus, and as Kützing imagined that the species discovered by him was an Alga, it was placed in the genus in which it seemed to have the nearest relatives.

2. *Cladothrix dichotoma.*—In Zopf's " Zur Morphologie der Spaltpflanzen " (1882) organisms which were indubitably Leptothrix ochracea were described by this writer under the name Cladothrix dichotoma owing to a false impression as to their identity.

3. *Chlamydothrix ochracea.*—Migula has thought fit to bring Leptothrix ochracea into the fold encompassed by his genus Chlamydothrix. As will be shown below the change of nomenclature is open to challenge and there are weightier reasons for retaining the old generic name.

Structure.—With the aid of a good microscope the threads of Leptothrix can readily be identified. A photomicrograph of the organism is shown in Plate I, Fig. 1. This was taken from the deposit of a ferruginous stream. As to the deposits it must be borne in mind that they are not composed of *living* representatives of the organism ; the observer sees only the empty tubular, iron-encrusted sheaths of the dead organism. These are readily identified as the remains of Leptothrix ochracea, the yellow-red, sharply defined sheath being of a very distinctive nature. Whilst the majority of the threads are straight, some show slight curvatures (Plate I, Fig. 2). Occasionally, as seen in Plate I, Fig. 3, the thread is thrown into a number of undulations of regular wave-length. In the deposits the threads that come under observation are very seldom whole organisms. They are rather fragments, as can be observed from an examination of their ends, which appear as if they had been snapped across. This would doubtless happen during the process of deposition and penetration of the ferric hydroxide.

The appearance of the living organism is very characteristic. The thread is tubular, delimited externally by a delicate membrane which is normally much thinner than the dead membranes as found in the deposits. The ends are rounded off symmetrically. When grown in artificial cultures devoid of iron the membrane remains thin and delicate (Plate I, Fig. 3), and even in nature this thinness and delicacy is observable. Usually, however, the membrane increases somewhat in thickness and assumes the familiar ochre-coloured appearance. The average threads measure $1\frac{1}{2}\mu$ to 2μ in thickness when free from an external deposit of iron. When dead and caked with iron the thickness may extend to 3μ and more. The length of the threads may reach to 200μ, and even greater lengths have been observed. In artificial cultures (Plate I, Fig. 4) they sometimes elongate considerably and assume an appearance markedly different to that found in nature. The threads increase in length, but whether the growth is intercalary or terminal has not yet been determined; on à priori grounds the former is the more probable.

There is concurrently a very slight increase in thickness; this, however, must be very small and must cease long before the extension in length has been completed, for long and short threads are not appreciably different in thickness, although both are appreciably thicker than very young threads.

Of the internal structure of Leptothrix we know very little. By the ordinary methods of staining only a uniform homogeneous structure can be observed within the threads. As is the case in the vast majority of bacteria, the threads show no distinction of cytoplasm, nucleus, and vacuoles. Cell-division by fission is a common occurrence in growing threads. Whilst in some both ends are rounded off, in others only one end is so formed, the other end being angular as shown in Fig. 1. In others again both ends are angular (Fig. 2). Until the first division takes place a young thread is invariably rounded at both ends. The reason for this failure of the cut threads to round off the end at which division has taken place is probably connected with the growing

rigidity of the membrane at this stage, as the iron
deposition is by this time comparatively far advanced.
A transverse membrane is never formed, the organism
differing in this respect from the genus Bacillus.   The

FIG. 1.   FIG. 1a.   FIG. 2.

FIG. 1.—Leptothrix ochracea.   Figure from Molisch's " Eisen-bakterien ".

FIG. 1a.—Leptothrix ochracea.   Thread is rounded at one end, square cut at
        the other.
FIG. 2.—Leptothrix ochracea.   Thread square cut at both ends.

details of the operation of division have not yet been
satisfactorily worked out.   In some cases I have
observed the formation of thickened rings of the
form shown in Fig. 3, at the point of division, and it
would appear that fission takes place between the
rings.   It is curious that the rings should often be

placed not transversely but obliquely to the membrane, and this accounts for the fact that in many of the older threads the surface of the ends is not set at right angles to the sides, but is obliquely placed. The shape of the thread under the circumstances is not unlike that of a quill tooth-pick. For the observation of living threads search must be made in the stream itself and not on the bed. Fluffy bits of floating or immersed material must be sought for, and these are best found in early summer or late autumn. Apparently at these times the constitution of the organic matter in the stream best favours the growth of the iron-bacteria. Threads that are covered with ferric hydroxide should be treated with dilute hydrochloric acid to get rid of this deposit. When this is done the membrane will be uncovered and will be found generally to be of a slightly yellow-red tinge and sharply delimited (Plate I, Fig. 1). It often happens that on treatment with hydrochloric acid the deposit peels off in segments, in which case the membrane is laid bare in some parts of the thread, whilst in the other parts the membrane is overlaid by the deposit (Fig. 4). In such cases it is possible to ascertain the fact that the deposit of iron in the old threads is *inside* as well as outside the membrane, so that the ferric hydroxide is actually inside the cell in the older threads, a fact which we must bear in mind when dealing with the physiology of the iron-bacteria.

Another point to observe in the structure of Leptothrix is that, in common with practically all low

organisms of this grade, the real membrane is invested
by a very tenuous, mucilaginous, under ordinary
circumstances invisible mantle. This envelops the
membrane and covers its surface much in the same

FIG. 3.

FIG. 4.

FIG. 5.

FIG. 3.—Leptothrix ochracea. Showing thickened rings between which
   division appears to take place. (Diagrammatic.)

FIG. 4.—Leptothrix ochracea. Portion of old thread, showing membrane
   from which the ferric hydroxide has been removed in parts: shows
   relationship between thickness of membrane and thickness of
   deposit of ferric hydroxide. m = membrane from which deposit
   removed.

FIG. 5.—Leptothrix ochracea. Appearance presented by old thread laden
   with thick deposit of ferric hydroxide. Outside surface rugged
   and encrusted with foreign particles.

way that a soapy froth covers the surface of the water
out of which it has been formed. The presence of
this mucilaginous layer has a considerable influence
in determining the final shape of the plant, for the
iron compounds drawn from the water are retained in
it, and undergoing solidification by oxidation, impart

solidity and opaqueness to an otherwise very tenuous and invisible structure. In the oldest iron-encrusted threads the membrane is often no longer recognisable as a separate entity, the strain on such a delicate structure having evidently become too great for its stability, with the result that it has broken at many points, and whilst the cylindrical structure is as a whole maintained, the organism appears as a roughly cylindrical mass of ferric hydroxide with a very irregular surface, often dotted with foreign particles (Fig. 5). Seen by itself, apart from its surroundings, such a mass would raise doubts as to its derivation from any organism.

According to Migula (1) Leptothrix is composed of a sheath within which a number of cells are lodged. This interpretation has not been confirmed by any other investigator who has studied the threads of Leptothrix. It is probable that Migula has mistaken threads of Cladothrix dichotoma for those of Leptothrix ochracea. The threads of the former, unless stained, appear to consist, as is shown later, each of a non-cellular tube, and it is only when treated with a reagent like iodine that the cells become evident. As young Leptothrix and young Cladothrix dichotoma threads are superficially not unlike, and as there has been no confirmation of Migula's statements on this point, the probability is great that the interpretation given above is the correct one.

The change, therefore, of the name Leptothrix ochracea to Chlamydothrix ochracea, instituted by

Migula on the strength of this supposed cellular structure of the organism, is not justified, and there seems no necessity for discarding the generic name Leptothrix which has been applied to this organism for the last seventy-five years.

1. *Multiplication.*—The most primitive method of multiplication, viz. one by fragmentation, is exhibited by Leptothrix ochracea. A portion of a thread is broken off, the two parts elongating once more after separation : each part then repeats the process so that in a short time a very large number of organisms is formed. In Molisch's artificial cultures (2), the threads were evidently often of great length (Plate I, Fig. 4). This is seldom the case in nature. Among the lower bacteria the formation of long threads is an indication of the intervention of unfavourable conditions, as the organism has apparently the strength to elongate, but not sufficient to enable it to divide. Speaking generally, in artificial cultures of bacteria short rods are found in the most favourable cultures, and long rods intervene only when conditions are unfavourable for growth. Although some long threads may be seen in natural waters, only comparatively short rods are normally encountered.

1*a.* According to Molisch an interesting mode of reproduction is exhibited when a colony (in an artificial plate culture) of this organism is pressed down gently with a coverslip. Numerous rods escape, develop motility, and then swim away. He does not appear to have investigated either the method of separation or the cilia of these motile rods. The probable presence of cilia may be inferred from the motility.

The chief interest lies in the fact that the rods assume motility, for the young threads observed in natural waters are never in a motile condition, so far as they have hitherto been observed. According to the same writer the motile rods sometimes attach themselves to long threads of Leptothrix, giving the latter a false appearance of branching.

2. *Reproduction by the Formation of Conidia.*— This mode of reproduction was first observed by Migula (1) who states that their germination takes place on the threads. I have repeated Migula's observations and followed the various stages of development. The process begins by the formation on the thread of numerous small protuberances (Fig. 6). Each is approximately 1μ[1] in thickness. Elongation takes place until the protuberance is about $1\cdot25\mu$ in length ; then constriction at the base follows, with the result that a small oval structure is cut off. In some cases the production of conidia is so prolific that the organism itself is completely buried in them. The subsequent development of the conidium is probably accomplished by its elongation and subsequent separation from the parent thread. This is rendered probable by the fact that in the case of some of them in their development the process of constriction is postponed until the length of the developing conidium is considerably greater than its thickness (Fig. 6 at *a*).

---

[1] μ (the Greek letter *mu*) is the unit of length universally used in Bacteriology, and $= \frac{1}{1000}$ millimetre $= 0\cdot00004$ inch.

The appearance of the conidium is of a very unmistakable character, and we find similar structures in Gallionella and in Spirophyllum. It is therefore surprising to understand the attitude taken up by Molisch (2). On page 43 of his " Eisen-bakterien " we read : " Die conidien von denen Ellis spricht, habe ich nicht auffinden können . . . mir will scheinen, dass aufgelagerte Bakterien zellen eine conidienbildung vorgetäuscht haben, was ja bei einer Rohkultur nicht unmöglich wäre ".

FIG. 6.

FIG. 7.

FIG. 6.—Leptothrix ochracea. Devolopment of numerous small protuberances on thread. These may either be cut off when quite short —being then known as conidia—or they may develop further before abstriction takes place. The latter method of formation is indicated in the lower part of the figure.

FIG. 7.—Leptothrix ochracea. A further development of the same nature as is shown in Fig. 6. The protuberances have at this phase become some quill-like in appearance.

It is absurd to imagine that a Leptothrix thread in a natural water would be covered with bacterial cells derived from other sources than itself. If the structures in question are bacterial cells it is more reasonable to assume that they are "bacterial cells" derived from the Leptothrix thread ; in other words, that they are of the nature of conidia. This is a far more probable explanation than the one which presupposes Leptothrix to have drawn together "bacterial cells" from a variety of other organisms in the neighbourhood.

IDENTIFICATION OF LEPTOTHRIX OCHRACEA

The recognition of this species is easily accomplished by the microscopic examination of the deposit in ferruginous streams, when there should be no difficulty in recognising the empty ochre-coloured threads depicted in Plate I. When these have been identified the water should be searched for the living threads, and a search should also be made for slimy streamers in the water, which will also in all probability be found coloured with the ferric oxide. The living threads are normally non-motile, cylindrical, rounded at both ends, and varying in thickness from $1\frac{1}{2}\mu$ to 3μ or 4μ. Hitherto motility has not been observed in these threads in natural waters. The threads are naturally straight, but bent and often undulated forms are not very uncommon (Plate I, Fig. 2). The majority, however, show straight threads with sharply delimited membranes tinged with the ochre-coloured hydroxide of iron.

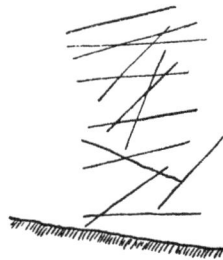

FIG. 8.—Leptothrix ochracea. Diagrammatic representation of a colony of Leptothrix threads. Threads are held together in the water by each one being in contact at one or two points with other threads.

I have occasionally observed a colony of Leptothrix in a natural water in which large numbers of threads were held together in such a way that one organism was in contact with one or even two others at single points, as is shown diagrammatically in Fig. 8. This peculiar disposition of the threads is due to the adhesion of the threads by their enveloping mucilaginous coverings, a disposition which must occasionally happen when a number of the organisms are growing in close proximity. When, therefore, Winogradsky (1) states that Leptothrix branches pseudo-dichotomously it is probable that a somewhat similar adhesion of individuals was observed by him, which conformed to a more regular pattern, producing the effect of a dichotomous arrangement. It is certain, however, that this false dichotomy is not a *characteristic* feature of Leptothrix, and phenomena of this kind which are only occasional in their manifestations, are not to be relied upon for the recognition of Leptothrix.

Leptothrix Ochracea from the Standpoint of the Engineer

Although such a cosmopolitan organism, and although present in such quantities in the waters which form the gathering grounds for the supply of many towns, this organism does not play an active part in those sudden visitations of the iron-bacteria so much dreaded by the water engineer, nor in the slower growth of the same class of organisms in the conduit pipes which furnish urban districts with their water supply. The reproductive cells of Leptothrix must, however, be present in all such waters, and, as will be discussed later, the writer is of opinion that the more troublesome Gallionella and Spirophyllum are both pleomorphic phases in the life history of Leptothrix ochracea. If such be the case the comparatively harmless Leptothrix must be regarded with suspicion equalled to that bestowed on Gallionella and Spirophyllum. Therefore, it is important to the engineer to note that Leptothrix forms resting cells in the form of conidia. As shown above these minute oval structures are formed in countless numbers in ferruginous waters, and constitute the "germs" out of which new individuals arise. The question of the location of the resting cells of the iron-bacteria has often been discussed, and on a few occasions has formed the subject of an official inquiry. It may be stated that the conidia constitute, so far as is known at present, the only form of resting cells of Leptothrix. I am quite convinced that the thread itself does not as a whole become a resting cell, the so-called *arthrospore* of de Bary. This phenomenon does occur occasionally in the lower bacteria, but there is no evidence of the existence of a similar mode of reproduction in the case of Leptothrix. In all cases the threads multiply by fragmentation and then die out, but the conidia which they liberate in countless numbers persist and, under favourable circumstances, germinate to form new threads. I have not followed the germination of the conidium of Leptothrix, but have done so in the case of the conidia of Spirophyllum which are identical in practically every respect, and the facts that have been ascertained for the conidia of one organism are applicable to those of the other. The conidia arise in water, and, until desiccation occurs, they remain in the water. As the water flows on countless numbers of these minute specks are carried along with it — it would take a million million of them to fill a small thimble—and each is potentially capable of germination when the conditions become suitable. Exceptionally they will be found in the soil in the neighbourhood of ferruginous waters, but this will happen only when the water containing the conidia has evaporated, leaving the bed dry. Under such conditions the conidia will be taken up by gusts of wind and scattered over the neighbouring soil. The following facts have been elucidated from observations of the growth of Leptothrix in pure artificial culture.

1. The minimum, optimum, and maximum temperatures of growth are respectively 5° C., 23°-25° C., 40° C.

2. It grows in darkness, in diffuse daylight, and in spite of bright sunlight.

3. It grows better the greater the amount of oxygen in solution in the water.

Care must be exercised lest this organism be confused with Cladothrix dichotoma, for although the latter is cellular, the cells cannot be made out without careful staining, and, until stained, young threads of Cladothrix dichotoma can easily be mistaken for Leptothrix ochracea.

CHAPTER II

GALLIONELLA FERRUGINEA (*Ehrenberg*)

Syn. CLADOTHRIX FERRUGINÉA (*Migula*)
„ GLŒOTILA FERRUGINEA (*Kützing*)
„ GLŒOSPHÆRA FERRUGINEA (*Rabenhorst*)
„ SPIRULINA FERRUGINEA (*Kirchner*)

THIS organism is best considered immediately after Leptothrix ochracea, because in the open it is very seldom that Gallionella is found apart from Leptothrix; in closed tunnels, however, it often appears as the predominating organism of the iron-bacteria.

Distribution.—It is probable that in iron-waters this organism is as widely distributed as the preceding, although in vastly inferior numbers. Unless a search be made especially for it, its presence is apt to be unnoticed owing to the overwhelming superiority in numbers of Leptothrix ochracea. Its presence has been recorded in Austria, France, and Germany, and I have found it to be very widely distributed in Great Britain, although very seldom in large numbers. When, however, there is a slight alteration in the external environment, it may multiply exceedingly fast and assume the rôle of predominating organism. Thus Adler (1) states that in a sample of iron-water containing a mixture of iron-bacteria Gallionella can be made to predominate simply by leaving the sample at room-temperature for a few days. Of more practical importance is the fact that its predominance is in many cases one of the results of passing water through pipes where, naturally, the conditions of growth are different, the chief factor being the absence of light. Thus the figures in Campbell Brown's paper (1) indicate the presence of Gallionella or Spirophyllum, or probably both, in the various mains in which slime had collected. In the case of the Liverpool water supply, the growth of these organisms in the mains between the reservoir at Vyrnwy and the filter beds at Oswestry, had resulted in a diminution of the discharging capacity of the pipe by from 20 to 25 per cent within the period of a few years. The presence of Gallionella in ferruginous medicinal springs has been recorded by Adler. He found Gallionella in such waters more frequently than any of the other iron-bacteria, its presence in abundance being found in approximately 30 per cent of such waters. It seems as though the determining factor were a diminution in the intensity of the light, for whilst in the direct sunlight Leptothrix predominates, in the waters which well out of the ground or which are made to flow through pipes, Gallionella springs into a position of importance.

PLATE II

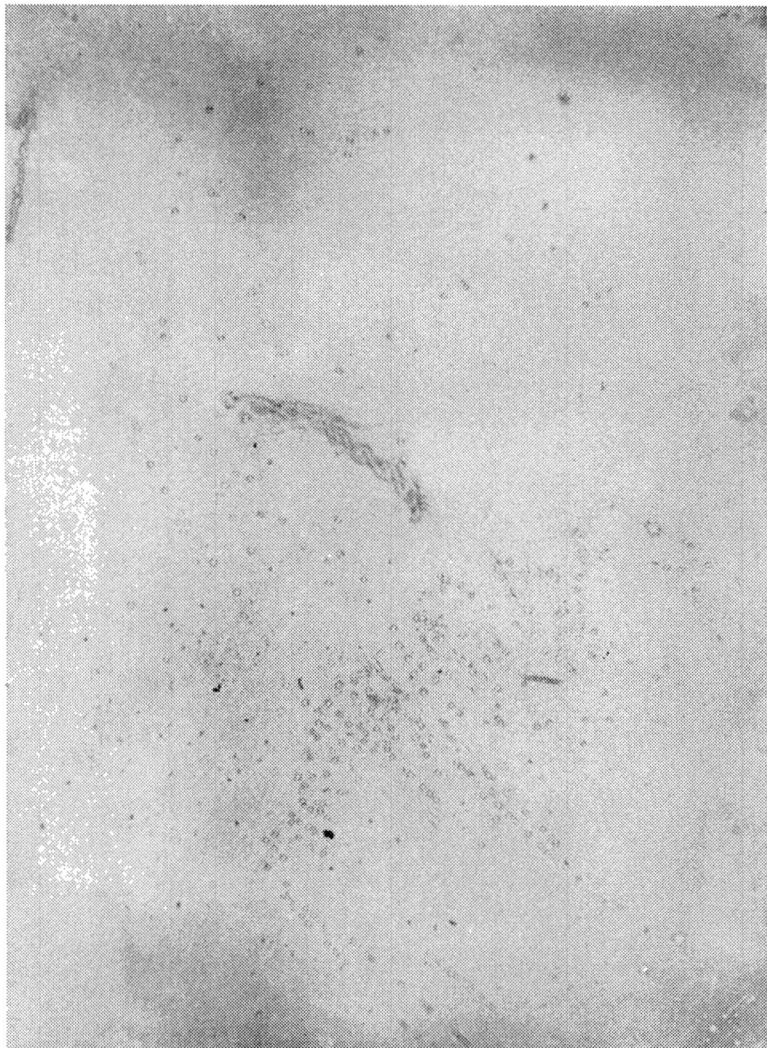

FIG. 1. × 350

GALLIONELLA FERRUGINEA

FROM BED OF FERRUGINOUS WATER

Structure.—We can gain a good idea of the structure of a typical Gallionella by twisting a hairpin spirally on itself. An example is shown in Fig. 9, and in Plate II, Fig. 1. The result of this winding is the production of a number of loops, and the two ends of the thread will be found close together (Fig. 9), just as are the two ends of a hairpin. The aver-

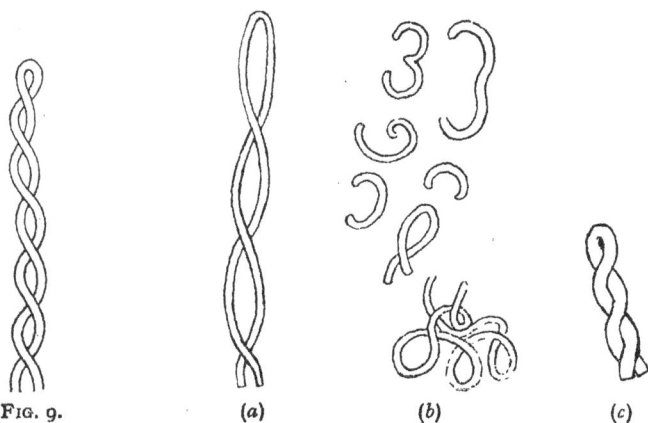

FIG. 9.   (a)   (b)   (c)

FIG. 10.

FIG. 9.—Gallionella ferruginea. Normal appearance of thread.
FIG. 10a.—Gallionella ferruginea. Showing thread somewhat loosely coiled.
FIG. 10b.—Gallionella ferruginea. Very young threads.
FIG. 10c.—Gallionella ferruginea. Tightly coiled thread.

age thickness of the thread is $\frac{1}{2}\mu$ to $\frac{3}{4}\mu$, although exceptionally it may reach to $1\frac{1}{2}\mu$. The loops formed by the twisted thread may be from 2μ to 6μ in amplitude. Again, a particular individual may form one loop, or several loops may result from its twisting habit. Examples of organisms showing variations in the number and amplitude of the loops are given in Fig. 10. Finally the two sides of the " hairpin " may not be of equal length, with the result that one stretches out far beyond the other (Fig. 11).

It can readily be imagined that when an organism

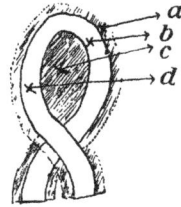

(a)   (b)

FIG. 11.   FIG. 12.

FIG. 11.—Gallionella ferruginea. Showing thread in which the first bend has not taken place in the middle of the thread. The two arms formed by this first bend are therefore unequally long, and when one side has become completely coiled, a large portion of the other and longer side is left over and remains in an uncoiled condition.

FIG. 12a.—Gallionella ferruginea. The bead-like appearance presented by many old iron-encrusted specimens. Surface is rough and uneven. The loops formed by the coiling thread have become filled with ferric hydroxide.

FIG. 12b.—Gallionella ferruginea. Diagrammatic representation of part of thread to explain origin of the bead-like structures shown in Fig. 12a.

        a = mucilaginous material on outer side of thread.

        b = lower arm of thread.

        c = mucilaginous material stretching across between the two arms b and d.

        d = upper arm of thread.

of this kind not only has this peculiar habit of coiling its body into loops, but also stores ferric hydroxide on its surface, we may expect it to exhibit a variety of forms. Further, it must be borne in mind that when coiling takes place the mucilage stretches across the loops formed by the coiling thread, and as iron is deposited on the thin films thus formed as well as on the thread itself, peculiarly shaped structures may arise. Instead of a coiling hairpin form, we often find mature Gallionella threads looking each like a string of beads, or a vibrating cord which is thrown into multiple loops. Some idea of the mode of formation of this particular form may be obtained by a reference to Fig. 12a. The interspace separating the two sides forming a loop becomes more and more solid as the iron deposit increases until finally a structure is produced which is very like Spirophyllum in appearance. We have already likened Gallionella in appearance to a cord set in multiple vibration. To understand the structure of Gallionella it must be remembered that the interspaces formed by the coiling threads are bridged by mucilage (see Fig. 12b). When this mucilage becomes impregnated with ferric oxide, it can readily be seen how the organism would assume the moniliform or bead-like structure, which is sometimes mentioned by engineers. Other names used by engineers are "the organism with a rosary form" or "the organism that looks like a row of beads". In the young stages, however, the twisted hairpin shape is apparent, the threads being yellowish-red in colour. It is not possible to make out anything of their internal structure. Migula (2), who has made an exhaustive examination of the developmental phases of Gallionella, states that whilst he was not able to distinguish a specific membrane, yet considered that there must be a very delicate covering bounding the exterior of the Gallionella thread, and his reason is conveyed in the following words : " Jedenfals muss aber eine etwa vorhandene Scheide, die ich wegen der äusserlich anhaftende Gelbfärbung durch Eisenoxydhydrat annehmen möchte sehr fein sein ".

He therefore *infers* the presence of a membrane but his reason cannot be said to be a good one as the deposition of a very fine layer of ferric hydroxide could very well take place on the outside of a thread devoid of a specific membranous outer layer. Hence the "äusserlich anhaftende Gelbfärbung" does not prove anything, for it could equally well be laid on a non-membranous as on a membranous thread. It is important that we should emphasise this point, for Migula only infers the presence of a membrane, which neither he nor others who have sought for it (Adler (1), Ellis (2)) can find. Yet it is on the strength of the presence of this membrane that Migula (2) has taken the responsibility of transferring Gallionella ferruginea into the genus Chlamydothrix (under the name Chlamydothrix ferruginea), the distinguishing feature of the members of which genus being the possession of an external membrane. In the author's experience no method of preparation or of staining has hitherto resulted in producing any differentiation of structure in these threads. Stripped of its ferric hydroxide all that can be ascertained in the thread is a homogeneous body of a semi-transparent consistency.

Migula distinguishes two kinds of Gallionella threads, one kind very delicate, irregularly wound threads, about 1μ thick, which sometimes lie alone, and sometimes unite to form a flocculent mass. The other kind shows very delicate threads which form chains.

As, however, all grades of transitional forms between these two kinds may sometimes be found in a Gallionella culture, they may therefore be regarded rather as the extreme forms of one kind of structure than as two distinguishably different types.

MULTIPLICATION

1. *Simple Fission.*—At all stages of growth a process of fragmentation takes place resulting in the liberation of small partly coiled fragments (Fig. 10*b*). These elongate and coil to form new organisms.

2. *Conidia Formation.*—I have ascertained that the conidia formation so characteristic of Leptothrix ochracea takes place also in Gallionella, though it is not such a frequent phenomenon as in the other organism (Ellis (2)). A typical case is seen in Fig. 13, which was taken from a culture that had been kept in diffuse daylight for seventeen days. It is interesting to observe that conidia formation appears to be accompanied by a loosening of the coils. In shape and

FIG. 13.— Gallionella ferruginea. Beginnings of conidia-formation.

manner of development the conidia are indistinguishable from those of Leptothrix ochracea.

In some instances it was noticeable that conidia formation on certain individuals had been so extensive that the organism itself had become completely hidden from view by the conidia, with the result that it had lost its characteristic bead-like appearance, and its form is roughly cylindrical. In this condition Gallionella cannot be distinguished from an old thread of Leptothrix.

THE CAUSE OF THE COILING OF THE THREADS

Detailed investigation of this phenomenon is lacking, owing chiefly to the want of success that has attended efforts to obtain artificial cultures of Gallionella. Certain conclusions may, however, be arrived at, based on comparisons with other organisms in which the same phenomena are shown. There seems little doubt that Gallionella offers another example of *contact-irritability*. The absence of a membrane doubtless enables the effect of irritability to take place with greater ease than if a membrane were present. It may be conjectured that contact with a particle in the water or with another organism brings about the first curvature, and then as the two ends in consequence approach contact is established between two different parts of the thread, with the result that the first twist-over takes place. Once established the process continues, each contact producing a further twining, this in turn producing contact at another

point.   In this way a state of matters analogous to
the "vicious circle" of the medical men is brought
about.   This process is continued until the deposition
of ferric hydroxide acting as a dead weight renders
further twining impossible, or possibly the cessation
may come about by the advance of "old age".   It
must be borne in mind that a power of movement
is inherent to protoplasm, and a long, naked thread
of protoplasm—for a young thread of Gallionella is
nothing else—is a peculiarly favourable structure for
the manifestation of movements, the stimulus for which
is supplied either by contact with a foreign body or
by one part of the thread coming in contact with
another part.

GALLIONELLA FERRUGINEA FROM THE STANDPOINT OF THE ENGINEER

As our knowledge of the iron-bacteria increases we
are able to make a juster estimate of the relative
importance of the various organisms in this class.
The importance to the engineer of Gallionella has
increased tenfold with greater knowledge.   As is well
known, incrustations in iron pipes begin as minute
dotted projections of a roughly limpet shape.   They
grow by the accretion of concentric layers, and ultimately coalesce.   Finally, unless cleaned out, the
pipes suffer an appreciable diminution in the diameter
of the bore.   In 1904 Campbell Brown (1) expressed
the opinion that while some of the nodular incrustations contained iron-bacteria, there were many cases

in which the incrustations were free from every trace of life, and he further laid down that the iron of the incrustation is in all cases derived from the iron of the pipe. Unfortunately, on the biological side, we must receive this investigator's conclusions with great caution. Judging from his plates and from his description, Gallionella (or Spirophyllum) is indicated as the organism concerned, and Schorler's subsequent research (1904) shows this inference to be correct. Schorler's results are very important and deserve careful consideration on the part of the engineer. He showed that the uppermost, easily washed away, layer of the incrustations which were found in the mains that supply Dresden contained threads of Gallionella in abundance. His most important contribution, however, was his demonstration of the fact that although the firmer parts of the incrustation showed no trace of Gallionella, yet the six-sided plates of a crystal nature, which were abundant in these firmer parts, contained each a fragment of Gallionella. That is to say, the crystallisation of these six-sided plates has followed a similar course to that which has resulted in the formation of coal nodules in the coal, namely, the collection of inorganic matter round an organic particle. Gallionella, therefore, alive and dead, is a focus of attraction for ferric hydroxide. These crystal plates measure 33μ to 144μ in diameter. We may conclude that even although Gallionella is not found in the firmer parts of incrustations, yet if its threads are found in the superficial layer the inference is warranted that they are present also in the more solid interior of the incrustation, although hidden from immediate view inside crystallised plates. Such being the case this organism must be held responsible for a large part of the incrustations that appear in supply pipes, culverts, etc., where a gently flowing water harbours the conidia of Gallionella, and where the light is of a very low intensity. In the light of Schorler's researches we cannot subscribe to Campbell Brown's statement (1) that "the iron of the incrustation is in all cases derived from the iron of the pipe". Of course, the converse is also not true, that the iron of the incrustation is wholly and necessarily derived from the water. But we are at any rate in a position to affirm that Gallionella is an important factor in the formation of the majority of the incrustations that form in water pipes. The part which is played by Gallionella in the formation of the slimy streamers that appear in water channels is probably very small ; other organisms, to be considered presently, play here a more important rôle. In Campbell Brown's paper (1) one of the figures (Plate I, Fig. 20) designated "Jelly developing inside slime," shows an undoubted example of Gallionella, but the majority of the threads in this figure belong obviously to some other organism, probably Cladothrix dichotoma. We have, however, sufficient material to show that in the matter of the slime, as well as in the case of the incrustations, Gallionella is a causative organism. With regard to the method of dispersal, all the remarks which were made with regard to the dispersal of the conidia of Leptothrix ochracea apply with equal force to Gallionella ferruginea, for as the writer has shown (Ellis (2)) conidia formation is characteristic of the latter as well as of the former.

THE IDENTIFICATION OF GALLIONELLA FERRUGINEA

The tendency to form spirally wound threads is the chief characteristic to be relied upon in the process of identifying Gallionella. In the young condition minute threads about $1\frac{1}{2}\mu$ thick, that are slightly curved, may with certainty be regarded as Gallionella. If such be found the microscopic examination should be directed towards ascertaining whether forms showing slightly later stages of growth—e.g. one showing a single loop (Fig. 10b)—are present. Reliance should not be placed on the discovery of a single specimen, but rather on the "atmosphere" of the microscopic field, on the ascertainment of a number of forms all individually different in detail, but all possessing the same general structure, namely, thin threads spirally wound in different ways. The figures shown in Figs. 9-13, should help in the identification of the various stages of growth. It will be found more difficult to recognise the older threads of Gallionella, for, as already stated, the "loops" become filled up with ferric hydroxide, and structures such as that shown in Fig. 12, are presented to the eye. This is the so-called "rosary" or "bead" form of Gallionella. In this condition an appearance very

similar to Spirophyllum is assumed.   In fact, Prof.
Molisch (2) has expressed the opinion that Spirophyllum is identical with Gallionella, obviously basing
the statement on the close similarity of the old iron-
encrusted forms of both species, and totally ignoring the
wide dissimilarity of the two organisms in their young
stages.   We shall return to this subject again after we
have described Spirophyllum.   The only safe way of
distinguishing the two species is to take advantage of
the fact that a microscopic field with iron-bacteria in
it will probably show these organisms at different
stages of growth.   If an examination shows a series
of stages similar to those shown in Figs. 9-13, Gallionella is indicated.   When a "bead" form is encountered it must be regarded as either Gallionella
or Spirophyllum.   It is only the younger forms of
growth that can be used for distinguishing between
these two species.

CHAPTER III

SPIROPHYLLUM FERRUGINEUM (*Ellis*)

*I*NTRODUCTION. — A new species of iron-bacteria was found by the writer in the neighbourhood of Glasgow. It is characterised by the possession of a flat band-like structure, the band being normally twisted into a spiral form. A model of this organism could be made by obtaining a flat piece of elastic band—e.g. one ½ inch wide and 4 inches long—and, while holding one end fast, twisting the other end round so as to make the band assume a spiral form. Examples of Spirophyllum ferrugineum are given in Plate III, Fig. 1 and Fig. 14. The bands vary in width from 1μ in the youngest forms to about 6μ in the fully matured specimens. As in the case of the other iron-bacteria the colour varies from a grey tone to the familiar ochre colour characteristic of the beds of ferruginous streams.

Distribution.—I have found Spirophyllum to be universally distributed in the ferruginous waters of Great Britain from the north of Scotland to as far south as Kent, and as far west as Cardiganshire (Ellis (1)). Like Gallionella, in open waters Spirophyllum plays a very subordinate rôle, the chief organism in such

PLATE III

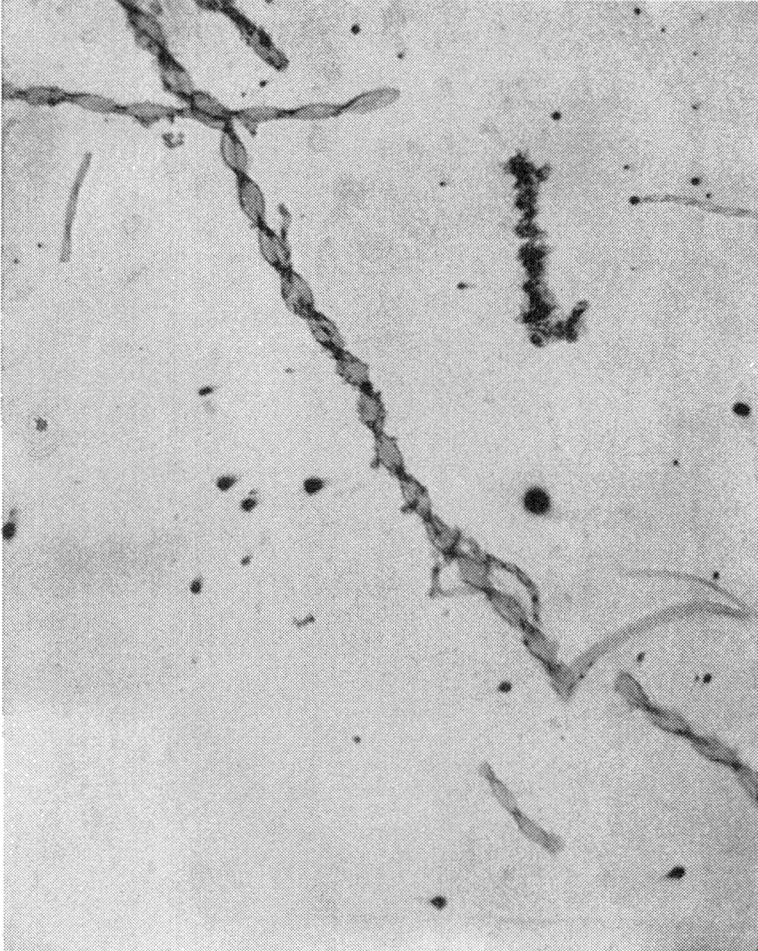

FIG. I. × 350

SPIROPHYLLUM FERRUGINEUM

FROM BED OF FERRUGINOUS WATER. SPECIMEN COVERED WITH CONIDIA

waters being Leptothrix ochracea. Occasionally, and here again it resembles Gallionella, it assumes the chief rôle, Leptothrix being either absent altogether or present only in a negligible quantity. The presence of Spirophyllum has been noted in America, and it

FIG. 15.

FIG. 16b.

FIG. 14.

FIG. 16a.

FIG. 14.—Spirophyllum ferrugineum. Fully developed band.
FIG. 15.—Spirophyllum ferrugineum. Tightly wound band.
FIG. 16a.—Spirophyllum ferrugineum. Loosely wound band.
FIG. 16b.—Spirophyllum ferrugineum. Two spirally wound bands in contact and commencing to wind round each other.

may safely be conjectured that it is as cosmopolitan in its distribution as the better known Leptothrix ochracea.

General Structure.—The smallest specimens measure only 1μ in width and about 2μ to 3μ in length. Even in these the spiral twist is manifest. One of the largest bands seen is shown in Plate III ; such individuals measure 6μ in width, and may extend to a length of 200μ or more. There is much variety in the number of spiral twists. In Plate III a large number is shown. In some the twining process has proceeded so far that the organism looks like a small portion of rope (Fig. 15). In others only one or two incomplete turns have taken place in the band (Fig. 16). The different varieties are shown in Figs. 16, 17. Occasionally two bands are seen in contact. In such cases not only is each band spirally twisted on itself, but each is further partially or completely twisted round the other Figs. 16a, 18.

The thickness of the bands is not uniform, the edge being thicker than the inner part. The same peculiarity was found by Warming to be characteristic of the members of the genus Spiromonas. In very young individuals this peripheral thickening is not evident. Its appearance in the older forms obviously affords a measure of protection to the organism, and to the need of such protection we may biologically account for its presence. The thickening at the edges is diagrammatically represented in Fig. 19.

Even when stained no differentiation of cell-structure can be distinguished, the bands being in this respect similar to the threads of Gallionella. Another peculiar feature is the appearance of the ends of the band which are all more or less square cut and never rounded. Even in the very youngest specimens of Spirophyllum in which a definite form begins to emerge, the ends are never rounded off.   Thus, as shown in Fig. 23, in which are shown young individuals the width of which are not greater than

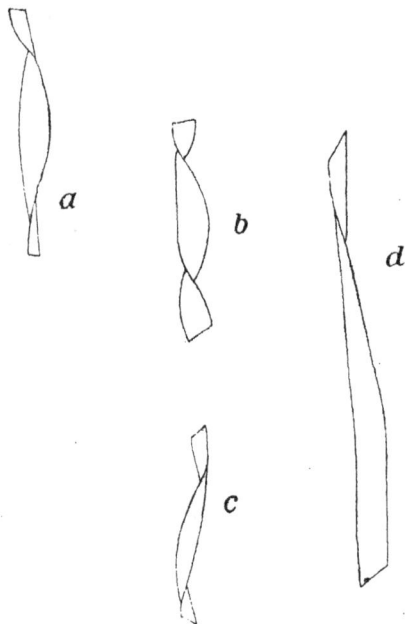

FIG. 17.

FIG. 18.

FIG. 19.

FIG. 17.—Spirophyllum ferrugineum.   Young bands at various stages of their growth.   The band shown at *d* has one twist : the others each two twists (× 800).

FIG. 18.—Spirophyllum ferrugineum. Shows the intertwining of two individuals, each of which is also spirally twisted on itself (× 800).

FIG. 19.—Spirophyllum ferrugineum.   Diagrammatic representation of section of Spirophyllum showing its thickened borders.

$1\frac{1}{2}\mu$, and in which cell division had certainly not begun, the ends are as sharply cut as they are in the mature organism.

Reproduction

1. *Fission.*—I have not observed the bands in a state of active fission, but their appearance and the relative positions of many of the contiguous bands leave no doubt that multiplication by fission normally takes place.

2. *Conidia Formation.*—In some cases abundant formation of conidia was observed. Slight protrusions appear on the surface of the band at a period of growth when the organism has not yet assumed the characteristic colour of the mature form. The method of growth and development of these conidia is precisely the same as for the similar structures of Leptothrix and of Gallionella. As is the case in these two organisms the majority of the conidia remain adherent to the parent organism. An example of an individual showing conidia on its surface is shown in Fig. 20.

I have been successful in obtaining active natural cultures in which the conidia were in process of germination, and in which, therefore, all stages of growth were exhibited, from the very earliest appearance of the organism after germination to the fully matured forms. The earliest phase that was observed is shown in Fig. 21, and is barely 3μ in length. The thicker U-shaped margin on the left of the organism may indicate the remains of the conidium wall, or it may indicate a portion of the organism which is not in the same plane as the remainder. It would be difficult to speak with certainty with regard to this small object had it stood alone, but the case is different when it is linked up by many different stages of growth to individuals the identity of which with Spirophyllum there is no doubt

FIG. 21.   FIG. 22.

FIG. 23.

FIG. 20.

FIG. 20.—Spirophyllum ferrugineum.   Band in process of formation of conidia (× 800).

FIGS. 21-22.—Spirophyllum ferrugineum.   Very young specimens.   The flat band form with angular ends peculiar to this species is observable even at this stage of growth (× 800).

FIG. 23.—Spirophyllum ferrugineum.   Very young forms about 1¼ in. in width and approximately 12μ in length.   All are flattened structures with angular ends, and two of them exhibit the first spiral twist (× 800).

about.   Already even at this very early stage the organism is flat and somewhat angular at the ends. A slightly more advanced stage is shown in Fig. 22. From this stage to that shown in Fig. 23 is only

a step, and from the latter again intermediate grada-
tions connect up to the various phases shown in Fig.
17. At the stages represented in the last figure
there is already a perceptible deposit of iron on the
membranes.

Motility is characteristic of the bands during the
early phases of growth. The movement is in the
direction of the long axis, although, as is to be
expected in an organism of this shape, there are also
slight rotary movements. The forward and rotary
movements are very slight ones, but quite distinct
and markedly different from the Brownian movements
characteristic of the finer particles that are usually
found in the same microscopic field. I was not able
to demonstrate the presence of cilia, and it is probable
that movement is not brought about by the activity
of structures of this nature. Independent movement
can be observed only in the very youngest of the
organisms, for it is never observed in individuals
in which the faintest trace of colour—due to the
deposition of iron—is noticeable.

The Cause of the Spiral Winding of the Bands.—
It is highly probable that the cause of the winding
of the bands is to be traced to *contact-irritability*,
although pending the successful initiation of artificial
cultures this must remain a mere hypothesis. Considerable support is, however, lent to this view by the
fact that two organisms are frequently found partially
wound round each other (Fig. 18). It is probable
that in the case of Spirophyllum, as in that of Gal-

lionella, contact with foreign particles initiates the spiral movement.   Here again we are dealing with a membraneless mass of protoplasm, which we can readily imagine to be susceptible to movements initiated by the stimuli set up by contact with foreign bodies, or by the contact of one part of the organism with another part of itself.

SPIROPHYLLUM FERRUGINEUM FROM THE STANDPOINT OF THE ENGINEER

The very close similarity of the adult iron-encrusted forms of Gallionella to those of Spirophyllum which are in a similar condition, makes it necessary to consider whether many references in the past literature to Gallionella may not have been in reality references to a mixture of Gallionella and Spirophyllum, or even to Spirophyllum by itself.   Pending further knowledge we must place Spirophyllum in the same category as Gallionella in estimating the importance to the engineer and the chemist of a knowledge of its life history.

THE IDENTIFICATION OF SPIROPHYLLUM FERRUGINEUM

All ochre-coloured bead-like individuals found in ferruginous waters and possessing dimensions within the limits given in the above pages are either Gallionella or Spirophyllum.   To distinguish between these two recourse must be had to a search for younger phases of growth.   A spirally twisted *band* indicates Spirophyllum, whilst a spirally twisted *thread* shows the presence of Gallionella.

ARE LEPTOTHRIX OCHRACEA, GALLIONELLA FER-
    RUGINEA AND SPIROPHYLLUM FERRUGINEUM,
    PLEIOMORPHIC[1] FORMS OF ONE AND THE SAME
    ORGANISM ?

As far back as 1856 we find the question of the identity of Leptothrix with Gallionella raised by Mettenheimer (1). This writer was satisfied that they were one and the same organism. Again in later years (1892) Hansgirg combined Leptothrix and Gallionella into one species because he regarded Gallionella as a developmental phase of Leptothrix.

The more these organisms are studied the greater is the conviction forced upon one that the three names given above have been applied to different life-cycles of one and the same organism. These three forms are always found in association in nature, and their methods of reproduction by fragmentation and by conidia formation are identically the same. If this be the case, whatever the causes which determine the particular form that is to be assumed, they assert their influences at a very early stage, for the form of the adult is found in very young individuals. The conviction of pleiomorphism is strengthened by the facts that all three are iron-organisms, and that all three are found in nature only in one type of water. The difference in shape appears at first to be destructive

---

[1] *Pleiomorphic.*—An organism is said to be pleiomorphic when it shows more than one independent form in its life-cycle.

to this conclusion, but fundamentally there is very little difference between Leptothrix and Gallionella. Both are thread-like forms, with this difference, that, whereas Leptothrix tends to form stiff threads, Gallionella very early begins to twist and form a coil, with the result that an organism superficially very different from Leptothrix is found. Spirophyllum appears fundamentally different owing to its band-like structure, which, as stated above, is noticeable in very early stages after germination. Against this, however, must be set the whole combination of circumstances which link up Spirophyllum with the other two, an association which is very close in every respect except that of its outward appearance. Of the three forms artificial cultures have been obtained only of Leptothrix ochracea ; and in these no traces have been observed of any pleiomorphic manifestations in the direction of Gallionella or Spirophyllum. This, however, is not conclusive, for, after all, in artificial cultures we have only one set of conditions, and it is possible that the absence of pleiomorphic forms in such may be due to the fact that the conditions are of a static nature, and are such as are favourable for the conservation of the Leptothrix form. It is obvious that we cannot reproduce in an artificial culture all the combinations of circumstances which exist in the *natural* waters in which Leptothrix lives.

It is suggestive that whilst Leptothrix should be the dominant organism in open shallow waters exposed to direct sunlight, Gallionella should predominate

in the darkened waters inside conduits, tunnels, etc.
This suggests that light may be a determining factor.
The verification of pleiomorphism for this group of
three organisms would, if ascertained, be a point
of considerable practical importance.   For example,
although Leptothrix is practically never found in
aqueducts, piping, etc., where the intensity of the
light is very small, it would be a mistake to imagine
that because it was found in the open waters leading
up to the aqueducts, etc., it was not necessary to
consider it in the light of an enemy.   If we knew that
it led a kind of Dr. Jekyll and Mr. Hyde existence,
the presence of Dr. Jekyll in the open is fraught with
danger if we consider his possible transformation into
the villain Mr. Hyde when he finds himself in the
darkened waters of the conduits.   Up to the present,
however, not a single proof has appeared to shed light
on the situation, and the problem still remains in the
realm of hypothesis.

CHAPTER IV

CRENOTHRIX POLYSPORA (*Cohn*)

Syn. LEPTOTHRIX KÜHNIANA (*Rabenhorst*)
   ,,   CRENOTHRIX KÜHNIANA (*Zopf*)
   ,,   CRENOTHRIX MANGANIFERA (*Jackson*)
   ,,   LEPTOTHRIX MEYERI (*Ellis*)
   ,,   MEGALOTHRIX DISCOPHORA (*Schwers*)

THIS organism is by far the most dreaded of the iron-bacteria on account of the suddenness of its attacks. It is the one which in consequence has received the greatest amount of attention. On the continent it is known as the " Brunnen-pest ". Probably it is normally present in all open water-storage reservoirs. Its growth and multiplication does not, and need not, under normal circumstances occasion any uneasiness, for, even if it excreted a virulent poison, the amount of this would be so small in relation to the vast amount of water into which it is liberated, that no danger need be apprehended. On occasions, however, the organism achieves what may be called "*calamity-growth*," when multiplication is so rapid that its results are visible to the naked eye in the form of a transformation of the normal colour of the water into a reddish-brown colour. A difference in the colour of the water may be then seen even in the small amount that is contained in an ordinary bath. The colour is seen to be due to countless multitudes of solid specks of a reddish-brown colour composed chiefly of Crenothrix polyspora. These specks are coated and permeated with a ferruginous compound. As may be expected the taste of the water during this period is not pleasant, neither is it pleasant to look at, for it is apt to lose its transparent pellucid appearance and assume a turbid and opaque character. Let it be stated at once, however, that no cases are known in which a person has died from the effects of drinking water contaminated by Crenothrix polyspora, or indeed by any of the iron-bacteria. The reason for this is not far to find. A water contaminated by, for example, the typhoid bacillus does not change its appearance, neither does the germ impart any taste to the water, and so it may be partaken with disastrous effects to the unsuspecting consumers. It is different with Crenothrix polyspora, for its presence in quantity is at once advertised by a change in the appearance of the water, and, further, we have not in the case of Crenothrix to fear the excretion of a virulent poison. Finally, even if it excreted a poison of this nature, we should still be protected, for Crenothrix cannot multiply inside our bodies, and there would be no more poison developed other than that which was in the water itself: there would not be a further accretion of poison inside our bodies resulting from the multiplication of the organism. The con-

PLATE IV

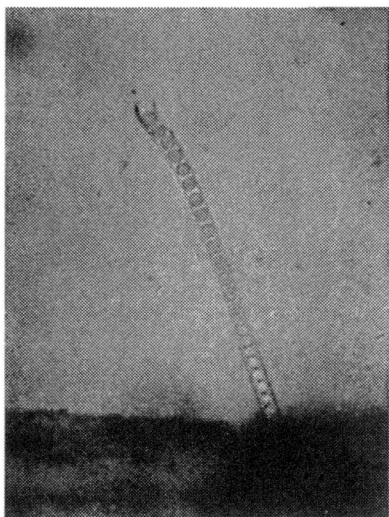

FIG. 1. × 400

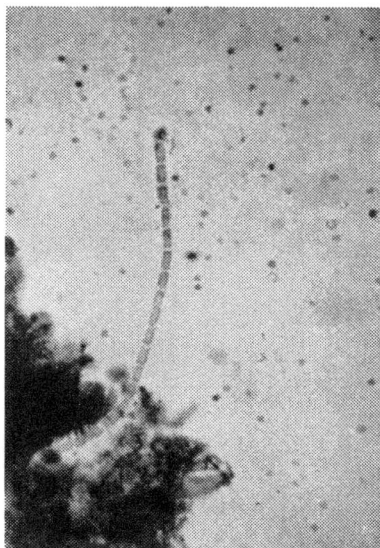

FIG. 2. × 400

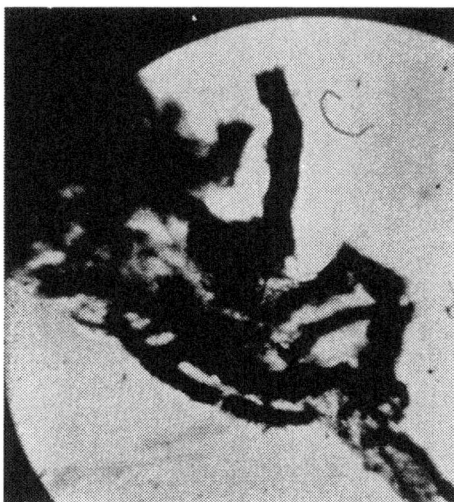

FIG. 3. × 400

CRENOTHRIX POLYSPORA

FIG. 1.   FROM FERRUGINOUS WATER FLOWING INTO RIVER CLYDE. CONDITION SLIGHTLY OLDER THAN THAT DEPICTED IN FIG. 2. SHEATH IS VISIBLE AND OPEN AT THE TOP. CELLS IN A SINGLE ROW

FIG. 2.   SEMI-MATURE THREAD. THICKNESS 1/20 MILLIMETRE. CELLS VISIBLE WITHOUT TREATMENT. APEX CLOSED. CELLS IN SINGLE ROW

FIG. 3.   BASEL PARTS OF CRENOTHRIX POLYSPORA IN STATE OF DECOMPOSITION. DESCRIBED UNDER THE VARIOUS NAMES OF LEPTOTHRIX MEYERI AND MEGALOTHRIX DISCOPHORA

ditions are, therefore, vastly different to those which would prevail if we had partaken of water contaminated with Bacillus typhosus. Crenothrix must, therefore, be regarded primarily as a pest to our water engineers, for it is their business to supply us with clean water free from tastes that are obnoxious to our palates.

Distribution.—We have records of the appearance of this organism in Germany, Holland, Austria, Hungary, Russia, France, Italy, and Great Britain, so far as Europe is concerned. Outside Europe it is further known to occur in India, America, and Japan. In each case attention has been called to its presence by the production by it of a major or minor "calamity". Therefore we do not hesitate to ascribe to Crenothrix polyspora a world-wide distribution, and it is perhaps not too much to assert that no country in the world is free from it. It is found in brooks, in water-pipes and in drain-pipes, and, normally, in small quantities in open waters, where it competes with other organisms of a similar grade of evolution. In this country I have found it in the Clyde in the neighbourhood of Glasgow, but some difficulty was experienced in finding it in any open water of this kind. Multiplication sufficient to rouse Corporations into action took place at Breslau in 1870, at Berlin in 1878, at Lille in 1882, and at Rotterdam in 1888. In this country the severest visitation took place at Cheltenham in 1896, which it will be instructive to discuss in detail. According to Schorler (1) Crenothrix is constantly found in the Tolkowitzer water (which supplies the city of

Dresden), and he states that growth is found at its best during the month of April. Again Adler (1) mentions that Crenothrix is constantly found in the Prag drinking water when this is allowed to stand for some time. It has been found that a certain amount of impurity in the form of organic matter in solution is a condition of its growth. This question will be dealt with in a later chapter. Samples of iron-water from Cheltenham which were examined by the writer did not contain a trace of Crenothrix polyspora either in the water or in the deposit. This does not signify the total absence of the organism ; rather the probability was great that the resting spores were present, but until these germinate it is not possible to recognise them as belonging to Crenothrix polyspora. The same negative results followed from an examination of ferruginous water from the Rotterdam water supply. The splendid example of Crenothrix polyspora shown in Plate IV, Fig. 1, was obtained from the Clyde. Although highly ferruginous the water in this particular sample showed only a sparse growth of Crenothrix, or indeed of any other organism. It is evident from a consideration of the above facts that a patient search will reveal the presence of Crenothrix in a large percentage of ferruginous waters in every part of the world, but that its numbers in such waters are very restricted. Even in such places as Cheltenham and Rotterdam which have had disastrous visitations, the distribution of Crenothrix polyspora in normal times does not attain a greater intensity than at other places. The conditions which speed up the rate of multiplication of Crenothrix are to be found in changes in the environment favouring the growth of this particular organism. There is no reason to suspect that calamities owe their incidence to any specific infection of the water by the reproductive cells of Crenothrix. These changes in the environment are discussed below.

CHELTENHAM CASE OF 1896

This town obtains its water supply from the head-waters of the river Chelt. After entering a settling pool the water flows over a brick weir into an open reservoir. The chief point of interest to us is that a steep wooded slope bounds one side of the latter, the slope being intersected by numerous streamlets, the result being that part of the water which flows into the reservoir *passes over boggy ground.* The beds of these streamlets show the ochreous deposits typical of ferruginous waters. The water passes from the reservoir to sand filter beds, and from thence to the service mains. On 1st March, 1896, the water of a school swimming baths showed a red colour and assumed a turbid appearance. Soon after the water for the house supply assumed the same colour, and developed an offensive smell. In April, while the water in the settling pool was normal, that in the upper reservoir was brownish-red. At the sand filter beds not only was the water red and turbid but the surface-film[1] and the water emerging from the filter were similarly coloured.   Further, when the surface-film was removed, the water that issued from the piled up heaps of scrapped surface-film was dark red in colour, but dried black and smelt abominably.   On 2nd May the reservoir had become like a great horse-pond, although the water in the settling tank throughout the whole of this time had maintained its normal appearance.   After the middle of May, when the climax was reached, matters began to mend considerably ; so much so, that by the middle of June the normal state of affairs had once more been reached.

DESCRIPTION OF CRENOTHRIX POLYSPORA

Seen under the high power of a microscope, Crenothrix normally appears as a thin thread in the water attached at one end to some foreign object, for example, a particle of dirt or a stone granule.   Each thread when semi-mature consists of a tubular, more or less cylindrical, sheath filled with a single row of cells.   We could make a model of the organism, at this stage, by filling a very long test tube with a single row of marbles ; the test tube represents the sheath, and the marbles the cells contained in this sheath. A semi-mature thread of this kind is depicted in Plate IV, Fig. 2.   The organism here shown was

[1] *Surface-film.*—This is a gelatinous skin formed on the surface of the filter, made up of diatoms, bacteria, colloidal particles, and numerous other microscopic animals and plants.   It is the essential part of a sand filter and must be renewed every two or three weeks.

approximately $\frac{1}{200}$ millimetre in thickness and from $\frac{1}{6}$ to $\frac{1}{4}$ millimetre in length. (A millimetre = ·04 inch.) A glance at this Plate shows that the cells within the sheath are distinctly visible. This may be noted here, for the distinctness with which these cells are seen without undergoing the process of staining is the feature which should be utilised in distinguishing young Crenothrix threads from Cladothrix. In the latter the cells are not visible until they have been stained.

As development proceeds the organism loses its cylindrical appearance and becomes much wider at the top than at its point of attachment.

Concurrently the cells in the upper part divide in three directions of space, with the result that at the apex of the thread we find not a single row, but a more or less irregular mass of cells. The appearance of a thread of this nature, now in its mature condition, is shown in Fig. 24. There can be little doubt that the widening of the sheath at the top is caused by the outward pressure exercised by the growing cells. In the fully matured condition the length of the thread extends to any size up to 3 millimetres, an enormous length for an organism of this class, whilst the thickness at the base varies from 3μ[1] to 6μ, and at the apex from 6μ to 8μ. We have described the organism in the semi-mature and in the mature condition. In the very young condition, in addition to being smaller

[1] μ (the Greek letter *mu*) is the unit of length universally used in Bacteriology, and $= \frac{1}{1000}$ millimetre $= 0·00004$ inch.

FIG. 25.

FIG. 24.—Crenothrix poly-
spora. Appearance pre-
sented by mature speci-
men (× 500).

    a = so - called " micro-
gonidia ".

    b = sheath, trumpet-
shaped in mature
forms.

    c = basal cells.

FIG. 25.—Crenothrix poly-
spora. Colony of young
threads (× 500).

FIG. 26.—Crenothrix poly-
spora. Semi-mature
thread (× 400).

    a = sheath, hollow and
cylindrical.

    b = cells.

    c = escaping cell, com-
monly called macro-
gonidium.

FIG. 24.

FIG. 26.

than the form depicted in Plate IV, Fig 1, the sheath
is *closed* (Plate IV, Fig. 2), not open, at the apex.
The young forms may sometimes be found in their
habitat in little clusters, each cluster being composed
of a dozen or more tiny threads, all arising more or
less in the form of a rosette, from a common centre.
Even at this stage the cells are distinctly visible without staining.

Development.—We shall now describe the course
of development. It must be borne in mind that the
stages here depicted are those followed by the organism in quiescent times. It will be explained afterwards
how it departs from this course resulting in the dis-
coloration of the water in which it grows. All plants
and all animals begin life as a single cell, and Crenothrix is no exception. In Fig. 24 spherical cells are
seen being extruded from the top, and in Fig. 25
cells are observed being cast out of an individual
containing only a single layer of these bodies. Each
one of the extruded cells is potentially capable of
developing to form a new filament of Crenothrix.
The cell divides into two, the process of division
being the same as that which holds when the bacillus
divides. Each half elongates, and so we have two
fully formed cells instead of the one we started from.
The process is repeated in precisely the same way,
with the result that four cells all in one row are formed.
By a continuation of this process a filament consisting
of one row of cells is produced (Plate IV, Figs. 1 and
2). Now each cell in this row has its own membrane

4

or protecting covering, and in addition the whole filament is enveloped by a mucilaginous sheath, which
cannot be discerned except by special staining.   It
is this sheath which keeps the cells together and
prevents them from breaking away.   This extra
mucilaginous envelope is present in most of the lower
micro-organisms.   The special feature of this structure in the case of Crenothrix polyspora is that it
subsequently hardens to form the sheath (Fig. 24*b*).
Hansen records a similar process of hardening of this
mucilaginous envelope in the case of yeast cells.   If
now we imagine the cells of a Crenothrix filament
to be taken out, the hardened sheath will present
the appearance shown in Fig. 27.   It is tubular but
chambered by the formation of transverse walls.
Each cavity is the shape of and only slightly bigger
than the cell which it contains (Fig. 28).   These
cells continue to divide and elongate *whilst the sheath
is hardening*, and as one end of the filament is fixed
to the ground the pressure generated by the elongating cells is exerted upwards along the line of the
filament.   The result is that the upward moving
cells are not deterred in their growth by the thin
barriers presented by the transverse walls, and the
same pressure causes also a rupture at the apex.
The result is that the sheath becomes a tube such as
is shown in Fig. 29.   The transverse walls have
disappeared and the sheath is open at the top.   Inside it are living cells that are still dividing and
elongating, but they are not now in direct contact

with the sheath. Consequently the latter is fixed and further growth of the cells drives the top cells out of the sheath. This is the condition which is shown in Fig. 26. As can be seen by a glance at Fig. 24,

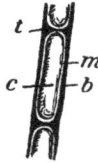

FIG. 27. FIG. 28. FIG. 29.

FIG. 27.—Crenothrix polyspora. Young thread. Shows cells invested in mucilaginous envelope which later hardens to a sheath (× 300).
     b = sheath.
     c = cells.
Membrane of cell not indicated.

FIG. 28.—Crenothrix polyspora. Portion showing cell slightly contracted (× 500).
     b = sheath.
     t = transverse wall of sheath.
     m = membrane of cell.
     c = cell.

FIG. 29.—Crenothrix polyspora. Shows tubular cylindrical sheath of half-developed thread. Transverse partitions of sheath have disappeared and a rent has been made at apex by the developing cells.

further development often—but not always—results in the cells dividing, not in one as hitherto, but in three directions of space. This change takes place only in regard to the cells at the top, and the result is that smaller spherical cells are formed and thrust out. At the same time lateral pressure is brought to bear on the sheath consequent on this new mode of division. It is this pressure which causes the upper part of the sheath to spread out in trumpet fashion. This is the ultimate stage in the life history of Crenothrix polyspora under normal circumstances. Further changes which take place are of a degenerative nature and indicate the beginning of the collapse of the whole organism.

It often happens that the organism retains throughout life the cylindrical form of its immature stage and that it does not assume the trumpet-shape customarily assumed by the adult organisms. Examples are shown in Plate IV, Figs. 1 and 2. Under these circumstances there is never more than a single row of cells inside the sheath, and the apical cells are similar in form, or approximately so, to the basal cells. The rupture of the sheath at the apex and the successive expulsion of the apical cells take place as in the trumpet-shaped form, only the cells thrown out by the cylindrical variety are larger than those of the trumpet-shaped kind, and are cylindrical, not spherical, in form. The large cylindrical expelled cells are usually designated *macrogonidia*, the name *microgonidia* being reserved for the small spherical ones expelled by the trumpet-shaped variety.

THE REPRODUCTION OF CRENOTHRIX UNDER NORMAL CONDITIONS

The reproduction of Crenothrix under normal circumstances differs considerably from the mode which prevails under those abnormal circumstances when multiplication takes place with such extreme rapidity. We have seen in the last section that individual cells are thrown out from the apex of the plant, and that these are of two kinds, namely, the larger cylindrical cells commonly called *macrogonidia*, and the smaller round cells, the so-called *microgonidia*. In the first place, it must be noted that there is no distinction of kind, only of degree, between these two varieties of cells; they are to be regarded as cells of the same kind, differing only in size and shape. In the second place, they are merely vegetative and not specially reproductive cells as their names would imply. They are simply cells that have been thrust out, and after expulsion they elongate to form new threads in a purely vegetative manner. The difference between them and the cells enclosed in the thread is a difference of "opportunity"—they have been liberated. In the third place, in an active colony of Crenothrix many intermediate grades in point of size between microgonidia and macrogonidia may be observed. They are to be regarded as cells which carry on the vegetative reproduction of the organism outside instead of inside the sheath. Of the large number thrown out not all form threads, and many either die or remain in a quiescent living condition in the beds of ferruginous streams. The living cells which undergo a period of rest constitute what de Bary would have denominated *arthrospores*. They differ in no single respect, so far as can be observed, from any other cell inside or outside the sheath. Under normal circumstances, reproduction is confined to the placid ejection of the microgonidia or macrogonidia from the apex of each filament.

THE REPRODUCTION OF CRENOTHRIX UNDER ABNORMAL CONDITIONS

It is evident from the above remarks that the procedure which we have outlined would not substantially alter the character of a body of water sufficiently to make the change perceptible to the eye ; the process is far too slow. An examination of the threads during an outbreak reveals a complete alteration, not so much in the character of the reproduction which is essentially the same, as in the speeding up of the output of cells. Zopf's well-known illustration,[1] taken during one of the German calamity growths, is one example of the appearance presented by a Crenothrix thread under such conditions. The difference between this condition and that described in the preceding paragraph is simply that the threads have speeded up their output of cells and that these cells not only surge out from the top, but also break through the sheath laterally and immediately commence to germinate. A number of these have not been able to liberate themselves completely from the sheath and have in consequence germinated in situ, with the result that in many cases the original thread

[1] See Lafar's "Technical Mycology," 1898, Vol. I, pp. 356 and 357.

is seen studded with dozens of streamers spreading out from its sides in all directions (Fig. 30).

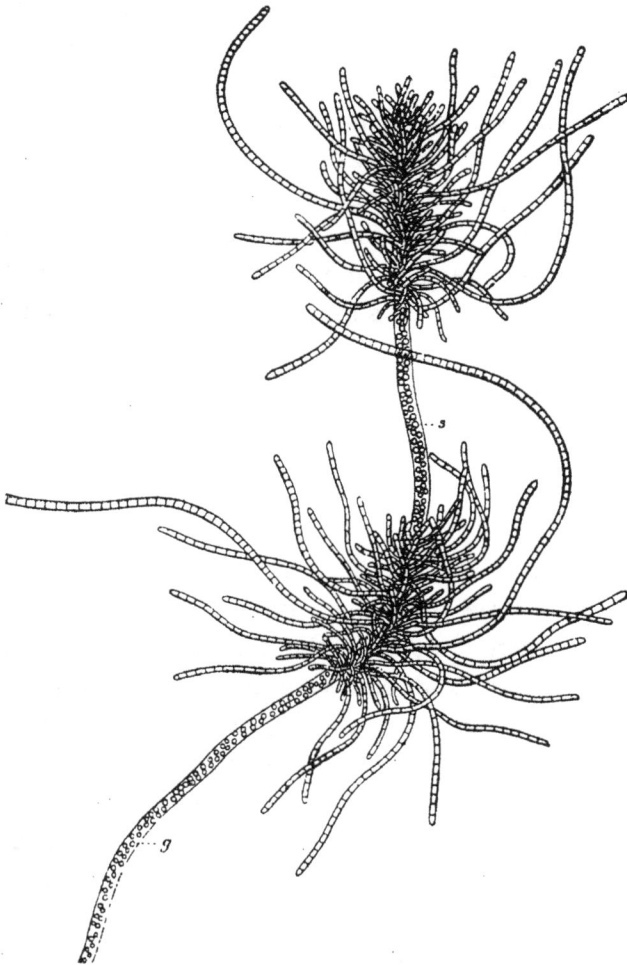

FIG. 30.—Crenothrix polyspora. Showing reproduction under abnormal circumstances. Cells germinate laterally through the sheath, so that immense numbers of fresh filaments are produced from each organism (× 600). (After Zopf.)

Eventually these streamers are liberated, and soon from them also cells well out from the apex and sides. In this way an inordinately large number of individuals is produced in a short time, and the effects of this development begin to be noticeable in a change in the colour of the water. Under these abnormal conditions a single thread produces at least a hundred reproductive cells, which probably *all* develop into threads at such times. Suppose a single thread to produce 100 reproductive cells in 24 hours, and suppose they all germinate. Then suppose the process to continue unchecked—

> After 1 day we have 10^2 filaments.
> ,, 2 days ,, 10^4 ,,
> ,, 3 ,, ,, 10^8 ,,
> ,, 4 ,, ,, 10^{16} ,,

After the fifth day we have a " horse-pond " ! That for a short time some such unchecked growth of all the reproductive cells of the threads takes place is evident from the suddenness of the Crenothrix attacks. Giard (1), for example, referring to the Lille outbreak, states that the calamity came upon them *suddenly*, and other accounts agree with this statement. Very interesting points about the growth of Crenothrix under these conditions were ascertained during the Cheltenham outbreak of 1896 (Garrett (1)). A microscopic examination of the water revealed a preponderating development of " cocci," as the small spherical microgonidia are often called (see Fig. 24).[1] They were small enough to exhibit Brownian movement so that the process of division of the cells which results in the formation of these microgonidia had been carried out to a far finer state of division than is usual in normal times. In fact, they could be described as micro-microgonidia. The products of this division, namely, the cocci, were found in profusion in the water and exhibiting motility,[2] some being oval, others round. In size they measured from 0.3μ to 1μ in diameter. Again, they were also found collected in Zoogloeae[3] (Fig. 31), of various dimensions, each coccus in the Zoogloea being brownish-red in colour. So far the statements accord with those made by the observers of similar visitations on the continent. Beyond this point Garrett's remarks must be received with caution. He refers to the fact that fresh filaments of Crenothrix develop inside the Zoogloea, and that a fresh filament of Crenothrix is derived, not from the sprouting or division and elongation of one of the cocci inside the Zoogloea, but

[1] The term "coccus" refers to very minute living spherical cells. It is oftenest applied to denominate the bacterial cells which are classed under the Coccaceæ, the spherical members of the group bacteria.

[2] Possibly Brownian, not independent movement.

[3] *Zoogloea*.—This is the name applied in Biology to a colony of cells held together by bulky mucilage to form either a membrane (e.g. a scum on a putrifying liquid) or a mucilaginous mass under the surface.

rather by rows of these cells arranging themselves into lines and thus forming new filaments. Such a procedure on the part of the cocci is so very far re-

moved from the behaviour of cells similar to these cocci and belonging to other organisms that we must re-ceive this statement with very great caution, especially in view of the fact that Garrett's work does not throw any light on the grounds which led him to come to his remarkable con-clusion. It is far more probable that the filaments seen by Garrett arose by the elongation and division of the cocci. Other interesting facts were brought to light. One was the breaking down of a filament at the ends and sides, and the emergence therefrom of a swarm of

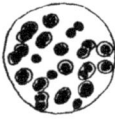

Fig. 31.—Crenothrix polyspora. Show-ing Zooglœa-formations. The cocci liberated from the Creno-thrix-filament col-lect in groups, be-ing held together inside a gelatinous envelope. Inside this envelope they multiply, produc-ing more cocci. Ultimately the cocci germinate to form threads. (After Zopf.)

cocci. This is a speeding up of the normal method of reproduction with a vengeance!

One can readily understand from these facts the rapidity with which a reservoir holding 84,000,000 gallons could become turbid and unfit to drink in a few weeks. Still another interesting point was the fact that the cocci continued to divide so as to repro-duce more cocci after leaving the filament and while enclosed in the Zooglœa. It was noticeable that in certain parts of the town the water was even worse in the pipes than in the reservoir, the reason for this being the increased multiplication by fission of the cocci, which took place in the pipes, *the division not being followed by the formation of filaments*, the cocci dividing to form minute fragments which developed into cocci. These cocci in turn gave rise to still more cocci, so that the pipes were filled with a vast multitude of forms, totally unlike the normal form of Crenothrix. In the writer's experience of bacteriological artificial cultures the manifestation of the phenomenon just mentioned has been found almost invariably to be the precursor of the destruction of the whole culture exhibiting it. The phenomenon of the production of minute structures, by, as it were, a frenzy of division, occurs occasionally in artificial cultures of both the Bacteriaceæ and the Coccaceæ, and is a sign of degeneration which usually puts an end to the sub-culture of the growth in which this occurs. It is probable that the above-mentioned peculiarity in the growth of Crenothrix is a phenomenon of a similar nature and precedes the collapse of the organism. From the point of view of the water-engineer the appearance of this phenomenon is to be regarded with satisfaction, for, although the immediate effects are not good, there is a promise in it of a period being set to the activity of the objectionable cocci.

FURTHER REMARKS ON THE NATURE OF THE SHEATH

As hitherto no mention has been made of the transverse walls of the sheath in any investigations on Crenothrix polyspora, a certain interest attaches to the mode by which the existence of these walls can be demonstrated. This is best done by the judicious application of a weak solution of iodine. This reagent renders the sheath sufficiently opaque to make it visible, and at the same time causes a slight contraction of the cells. The iodine must be applied at the edge of the coverslip and sucked through the water between the coverslip and the glass slide which

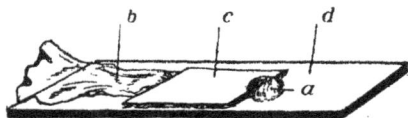

FIG. 32.—Crenothrix polyspora. To illustrate method of staining to secure best results with living material.

a = iodine.
b = blotting or filter-paper.
c = coverslip.
d = glass slide.

contains the Crenothrix, by placing a small bit of filter or blotting-paper on the opposite edge of the coverslip (Fig. 32). In this way the Crenothrix lies in a stream of iodine-laden water, and its gradual change of colour can be noted by keeping it under microscopic observation during the whole time that the stream passes over it. The change is not permanent, and so only a fleeting glance can be secured of these transverse walls. It is obvious that the experiment must be performed on young threads, such, for example, as are shown in Plate IV, Fig. 2.

MODE IN WHICH THE CELLS DIVIDE

The procedure of cell division is precisely the same as is seen among the lower bacteria, by the dividing bacillus, the main steps of which may be briefly enumerated :—

1. The cell elongates.

2. Very soon after the commencement of the elongation the middle of the elongating cell begins to show a constriction.

3. The elongation and the constriction proceed apace.

4. A transverse membrane is thrown across the dividing cell at the point of constriction, thus dividing it into two cells.

5. The transverse membrane, at first thin, gradually thickens and the two daughter cells are separated completely by a split through the middle of the transverse membrane. The split, of course, is parallel to, and not at right angles to the direction of the transverse membrane.

6. The daughter cells gradually draw apart and are completely separated.

MODIFICATIONS FROM THE NORMAL TYPE OF CRENOTHRIX POLYSPORA

In some waters very striking modifications are produced by the effects of the environment on the transparent sheath of Crenothrix polyspora. Thus Schorler (1) describes a modification which consists of long threads, 200μ-300μ long, of an almost worm-like appearance. This is doubtless the same type of Crenothrix which misled the present writer when he first observed it and led him to describe it as a new species under the name of *Leptothrix Meyeri* (Ellis (4)), which Schwers, under a similar error, has described under the name *Megalothrix discophora* (Schwers (2)). The sheath is very thick, mucilaginous, deep brown-red in colour, and somewhat wavy in outline. Seen apart it is totally unlike the form assumed by the normal Crenothrix polyspora, and not unlike a thread of *Leptothrix ochracea* with a very much swollen sheath. This modification is shown in Plate IV, Fig. 3. All doubts as to its identity with Crenothrix polyspora were set at rest by one of them being found by the present writer *in organic connection with Crenothrix polyspora*. It was seen that the "modification" was the basal portion of a normal Crenothrix thread, the sheath of which had become very much swollen and the cells of which had disappeared. Another modification is that which Giard (1) found to be the cause of the infection which took place in the water reservoirs of Lille in 1882. The organism was identified by him as the Crenothrix Kühniana of Rabenhorst and differed from Crenothrix polyspora in that the cells inside the sheath were motile.

To this class must also be reckoned the Crenothrix manganifera claimed by Jackson (1) to be a species quite distinct from Crenothrix polyspora. Inasmuch as the new species differs from Crenothrix polyspora only in the possession of a thicker sheath, we can scarcely subscribe to Jackson's contention that the two species are distinct. The thickness in Crenothrix manganifera is caused by an unusual amount of manganese on the sheath, this substance taking the place of iron as the predominating element. We know, however, that Crenothrix polyspora will take up manganese as readily as iron, and also that it shows a variation in thickness according to the nature of the medium in which it lives. There seems no reason for not regarding Crenothrix manganifera as an example of the effect produced on Crenothrix polyspora by living in a medium highly charged with assimilable manganese.

IDENTIFICATION OF CRENOTHRIX POLYSPORA IN THE MASS

The development of Crenothrix polyspora in large quantities is usually heralded by the appearance on the sides of the walls and at the bottom of the reservoir, if the latter be shallow, of greyish flocculent masses.   Minute specks of dirt in the water will be found to possess many streamers in attachment. These are, when young, of a white or greyish-white or slightly brown colour.   Schorler (1) reports that woolly-like loose masses of muddy consistency are found in the water when Crenothrix begins to multiply to any appreciable extent.   The signs in fact are the same as those which are associated with the excessive

multiplication of any one of the plant micro-organisms
that abound in such waters as Crenothrix lives in.
When the above manifestations are observed confirmation of the responsibility of Crenothrix must
therefore be sought for with the aid of the microscope, and the identity of Crenothrix established by
an examination of the small flocculent grey masses
that are so abundant in the water.   Again, according
to Schorler, when collected in a flask water infected
with Crenothrix forms a flocculent layer at the bottom,
white in colour.   This layer consists of threads which
contain no iron or manganese compounds on the surface of their sheaths.   As will be shown later, there
is no single member of the group of iron bacteria
concerning which it can be said that its growth is
necessarily associated with the presence of iron compounds in the water in which it abounds, and, it is
evident from the whiteness of these threads that during an outbreak the first rush of development is independent of the amount of iron in the water.   We
shall see later that a far more important factor is
the nature and amount of the organic matter that is
present.

In the Cheltenham outbreak an offensive odour
accompanied the growth of Crenothrix on the large
scale.   It is impossible at present to state the amount
of reliance that we can place on this factor as a
diagnostic feature, for there appears no mention of
any offensive odour in the accounts of the continental
outbreaks, and it is possible that in the Cheltenham

case other organisms had participated in the riot of growth. Some support is lent to this possibility by the statement made by Garrett (1) with regard to his liquid artificial cultures. Some of the reservoir water was placed in a beaker and to it was added a little of the red fluid drained from the sand at the bottom of the reservoir. In this medium the cocci multiplied in a Zoogloea condition and ultimately formed a deposit at the bottom of the beaker. From the cocci filaments were later produced, and during the process *the water changed its colour from red to green.* There seems no other way of accounting for this greenness except on the assumption of the presence of green algæ.

We may mention here the stages in the growth of Crenothrix as they were observed at Cheltenham in 1896 :—

1. April.—Complaints made as to odour, turbidity, and deposit in the water.

2. May 2nd.—Reservoir looked like a great brown horse-pond.

3. May 11th.—Reservoir assumed olive and purplish tints.

4. June 1st.—Reservoir assumed a dark olive-green colour.

5. June 5th.—Green predominated over red.

6. June 10th.—Water normal.

It is not difficult to understand the probable causes which operated to produce a change back to normal conditions :—

5

1. The growth of green algæ would cause the liberation of a large quantity of oxygen.

2. This oxygen would oxidise the organic matter in solution thus rendering it useless to Crenothrix as a source of food.

3. The Crenothrix would in consequence die from starvation, and the organic matter, the primary cause of the trouble, would disappear.

We shall return to the subject of the causes contributing to the growth of Crenothrix in Chap. VIII.

IDENTIFICATION OF CRENOTHRIX POLYSPORA UNDER THE MICROSCOPE

It must not be imagined that Crenothrix will always be encountered in nature in the form of the trumpet shape, tubular sheath usually depicted in the text books (Fig. 24). At the time when there is most need for an accurate diagnosis of water suspected of containing Crenothrix, the organism is either in the condition of thin immature threads (as in Fig. 26), or in the Zoogloea condition, consisting of a number of cocci contained inside a gelatinous mass (Fig. 31). In the latter case the cocci alone cannot be relied upon, for many other organisms form these bodies which often become embedded in small Zoogloea masses. In the former case identification can be ensured by its being borne in mind that the *cells* are visible in young threads of Crenothrix but not the sheath, whilst in young threads of Cladothrix dichotoma, which Crenothrix resembles most closely, the *sheath* is visible but not the contained cells until they have been stained. The only other organism with which it could be confused would be Beggiatoa alba at the initial stages of the disintegration of that organism. In order to be saved from confusion with this organism it must be borne in mind that the life histories of Crenothrix and Beggiatoa pursue courses which are entirely different. It happens that at a particular phase in its growth one of them superficially resembles the other when this other is at a particular phase of *its* growth.

It is therefore obvious that under such circumstances one must not place reliance on the appearance presented by any particular individual organism, but rather on the whole environment, and the appearance of other individuals in the same field must be taken into account. In a field of Beggiatoa alba individuals at a certain phase of development will show a superficial resemblance to certain individuals at a certain phase of development in a field of Crenothrix polyspora, but *regarded as a whole,* the two organisms pursue very different paths in their development, and consequently a bird's-eye view must be taken of the whole microscopic field. We may express our meaning symbolically. The letter "a" occurs in "man" and in "ape," but the two words are quite different in appearance. If we saw the letter "a," and knew that it belonged either to "man" or to "ape," we could settle the problem as to which word it belonged to only by seeing whether it was

connected with "m" and "n" or with "p" and "e".
Obviously we must know from previous experience
the appearance presented by Beggiatoa at the different
phases of its life, and we must know the same with
regard to Crenothrix. Armed with this knowledge
there is no difficulty in distinguishing a *field* of Crenothrix from one of Beggiatoa. Generally speaking,
Beggiatoa filaments are largely filled with round globules of sulphur which appear like small black circles
inside the filament. Crenothrix never possesses these
rings. Again, a healthy Beggiatoa is never divided
into cellular compartments ; the filaments of Crenothrix always show their cellular form. In a Beggiatoa
field many of the individuals will be in a motile condition ; this is never the case in a field of Crenothrix.

CAUSES WHICH ORIGINATE AN OUTBREAK OF CRENOTHRIX POLYSPORA

Here we are dealing with an organism which,
unlike the three others mentioned above, is apt to
break into rapid growth with a startling suddenness.
We have already stated that the chief factor in determining the growth of Crenothrix is the amount and
nature of the *organic matter in solution in the water.*
In the case of the Cheltenham outbreak we read that
a part of the water in the reservoir had, previously to
its inflow, passed over boggy ground. Further, we
are told that the sources of the river Chelt, the head-
waters of which supply Cheltenham, are fed partly by
the drainage of cultivated uplands. Lastly, the organic
matter was obviously present in a more concentrated

form after the drought which took place in the spring of 1896 and which caused a diminution in the volume of water in the upper reservoir. Turning to the Lille outbreak of 1882, Giard (1) tells us that industrial refuse had contaminated the water supply previous to the outbreak, and that in addition, the water had become contaminated by the water of a neighbouring bog. Again, Molisch (1) relates that the Prag drinking-water forms an admirable medium for the cultivation of iron-bacteria, and then he goes on to state that this water is *soft*, and that before its arrival in Prag it passes through a long stretch of *moorland* country. These three examples are sufficient to prepare us for the ample confirmation which has been derived from physiological experiments of the fact that when the organic matter in solution reaches a certain degree of concentration, other factors not being inimical, we may expect an outbreak of Crenothrix polyspora. It will be observed that no mention is made of the iron in solution in the water, nor of the widely spread idea of Winogradsky, which was never proved, that the growth of the iron-bacteria was dependent on the oxidation by them of ferrous into ferric compounds. We shall return to this subject more fully when we come to consider the physiology of these organisms.

In estimating the causes at work it is not necessary to emphasize the presence of reproductive cells of Crenothrix, for the incidence of the outbreak is not so much due to the appearance of the cells as to the appearance of the conditions which facilitate their growth. We may take it that practically all reservoirs in this country are more or less always infected by Crenothrix. They are in fact similar to our bodies, which, we are told, almost always harbour the germs of tuberculosis. It is only when the body offers the conditions of growth that Bacillus tuberculosis manages to make any headway. And the same applies to a water reservoir in relation to Crenothrix polyspora. If a stationary mass of water holds organic matter in solution, and especially if this organic matter is the result of contamination with bog-derived water, the chances of a calamitous growth of Crenothrix polyspora are greatly increased. Other contributory factors are an adequate supply of oxygen and the absence of serious competitors in the shape of other microorganisms. We know as yet very little concerning specific substances which act deleteriously on the growth of Crenothrix ; this is chiefly due to our failures to produce pure artificial cultures. According to Campbell Brown (1) the presence of bicarbonates, which imply the presence of lime and magnesia, seem to oppose rather than to accelerate the growth of Crenothrix. He states also that an appreciable quantity of iron in solution, in combination with organic matter of an acid character, is an invariable accompaniment of the production of the slime which is associated with the development of iron-bacteria. Finally, it is stated by the same writer that iron-bacteria never grow in an alkaline medium. These remarks were not made specifically with regard to Crenothrix polyspora, and we cannot verify their accuracy until artificial pure cultures have been instituted.

CHAPTER V

CLADOTHRIX DICHOTOMA (*Cohn*)

Syn. SPHÆROTILUS DICHOTOMUS (*Migula*)

„ CHLAMYDOTHRIX SIDEROPOUS (*Molisch*)

THIS interesting organism which was first described by Cohn, in 1875, consists of long thin threads usually attached at one end to various objects at the bottom and sides of stagnant streams. Like Crenothrix each thread is made up of a single row of more or less brick-shaped cells enclosed in a tubular sheath. Unlike Crenothrix the organism remains of uniform thickness even when the length attains the gigantic measurement of one millimetre. The cells divide only in one direction of space so that they never lie two deep in a thread. In long threads the cells exhibit a little variation in breadth, but not much, with the result that the thread along its whole length presents a very uniform appearance. As already stated, the individual cells are not distinguishable until they have been treated with a reagent that stains the cells but not the enveloping sheath. Iodine is by far the most efficient stain for demonstrating the fact that there are cells enclosed in the Cladothrix sheath,

as the cells take up the stain but not so the sheath
which remains transparent and colourless.

Each one of the enclosed rod-shaped cells measures
3μ to 6μ in length with a width of approximately 2μ.
The cells are shown in Plate V, Fig. 1, and on a

FIG. 33.   FIG. 34.

FIG. 33.—Cladothrix dichotoma. Cells shown somewhat contracted to enable the transverse septa of the sheath to be seen.

a = sheath.

b = cell.

c = transverse septum of sheath (× 1300).

FIG. 34.—Diagrammatic representation of true dichotomy.

much larger scale in Plate V, Fig. 2. As in Crenothrix polyspora the sheath forms transverse bars
between each cell, so that the sheath is made up of
a tube divided into compartments. Each compartment contains one cell. The relationship of cells
and sheath is represented diagrammatically in Fig. 33.

PLATE V

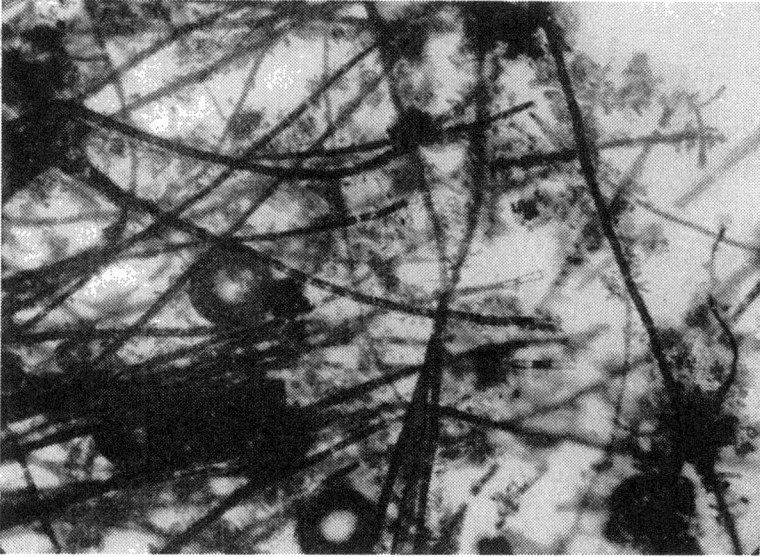

FIG. I.   X 400

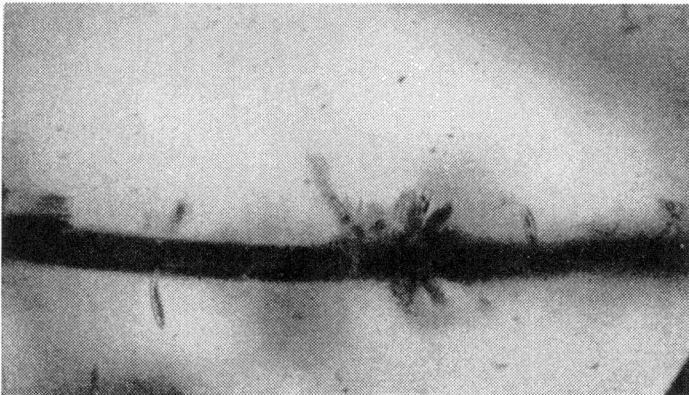

FIG. 2.   X 2500

CLADOTHRIX DICHOTOMA

FIG. I.   THREADS IN NATURAL CONDITION AND STAINED TO SHOW CELLS ENCLOSED
IN THE THREADS

FIG. 2.   PORTION OF THREAD SHOWN ON LARGER SCALE TO SHOW DIVISION OF THREADS
INTO CELLS

In most text-books this plant is reproduced showing false dichotomy. In true dichotomy the growing point of a plant divides into two parts each of which grows into a branch, and one or both may again bifurcate in the same way. The ultimate result is a

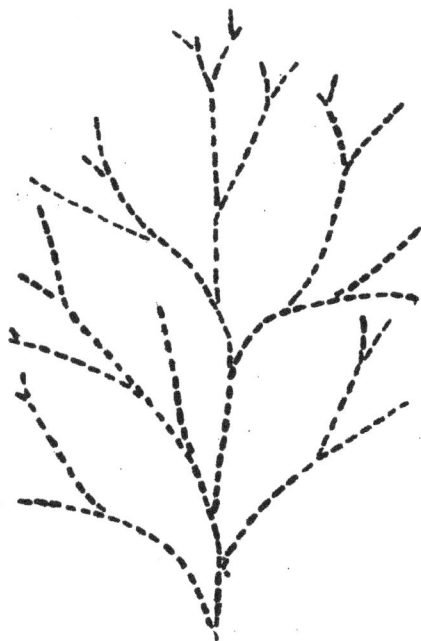

FIG. 35.—Cladothrix dichotoma. Diagrammatic representation of false dichotomy. For explanation see text.

system somewhat similar to that shown diagrammatically in Fig. 34. Under certain conditions this system of branching is *simulated* by Cladothrix dichotoma and a colony of this organism appears as is shown in Fig. 35. This condition is brought about in the following way. Normally the cells at the end of the thread slip out through the open mouth at the top of the sheath, but in addition some cells slip out *laterally*, cutting their way through the substance of the sheath. Usually these manage to get clear away, but occasionally they remain adherent to the sheath and there elongate to form a thread in attachment to the parent plant. This presents the appearance of a branch fixed to the thread, and when the process is repeated by several other cells in different parts of the thread, and also by the cells of the "branches," which therefore in their turn will have "branches" adherent to them, it can readily be imagined how a colony of threads is formed which has the general appearance shown in Fig. 35. The branching is thus seen to depend entirely on the adhesive capacity of the sheath, and it is clear that the connection between a filament and its branch is a purely fortuitous one; the branch has simply not succeeded in freeing itself from the parent, and in such a case the connection is mechanical, not organic. In the majority of cases, however, in which Cladothrix threads are encountered, they are not encumbered with any "branches". They appear as isolated long thin filaments, usually in great numbers, attached to a common object, very often a blade of grass or a particle of mud. The normal appearance of Cladothrix *in statu Naturæ* is shown in Plate V, Fig. 1. When single cells slip out laterally a compartment in the sheath is left empty. This space, however, is not allowed to remain unchanged, for it soon becomes filled with the same mucilaginous material of which the sheath is composed. At such points, therefore, there is a break in the continuity of the cells, the space at such interruption being occupied

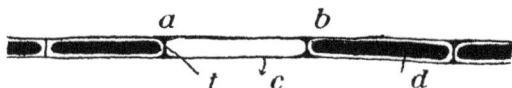

FIG. 36.—Cladothrix dichotoma. Portion of thread from which a cell has been removed laterally.

c = sheath.

d = cell.

t = transverse septum of sheath.

Between a and b is a compartment, lately occupied by a cell and now empty.

by a *broad* band of mucilage (Fig. 37). In course of time the growth of the cells diminishes the breadth of this band, until ultimately it becomes so thin that special treatment is necessary to demonstrate its presence. It is important to note that in the young

FIG. 37.—Cladothrix dichotoma. Portion of thread showing a gap between two cells, due to escape laterally of one of the cells. The space c becomes gradually filled with the sheath material. This in turn is gradually encroached upon by the developing cells, until the space at c is of same dimensions as those between the cells at d and at e.

threads every cell is separated from its neighbours on either side by a thin transverse bar of mucilage (Fig. 36). We shall return again to these transverse bands. The growth in length of the thread goes on always in one direction only, namely, along the length of the thread. In this way the number of cells in a thread may become very many. If the thread is attached by one end there may be as many as a hundred cells in a single filament. As the latter lengthens, however, a radical change is effected by the hardening of the sheath and the disappearance of the transverse bars which separate the cells. The cells still continue to divide and elongate, but their movements are circumscribed owing to the fact that they are enclosed in a hollow rigid tube. The sheath is no longer a clinging envelope of the cells, but a single-chambered hollow tube open at the top. This condition has been brought about by the pressure exerted by the elongating cells which has sufficed not only to break through the thin transverse membranes, but also to force an opening at the apex (Fig. 38). In order to explain the disappearance of the transverse septa and the eruption of the cells successively through the apex, two facts must be borne in mind.

1. That the sheath *gradually* hardens and becomes completely detached from the cells.

2. That while this process is taking place the cells do not cease to divide and elongate, consequently the resulting pressure is directed along the line of the thread.

The cells are now pushed outwards one by one from the apex. After a time growth of the cells ceases and the final stage is represented by individuals consisting each of a hollow tube with usually one or two cells inside it (Fig. 39). The process of sheath hardening is thus essentially the same as that which takes place in the formation of the sheath of Creno-thrix polyspora. In both cases there is the same gradual hardening of the sheath, whilst at the same time a passage to the outside is maintained by the cells by their continued growth and elongation.

Free Threads.—In addition to the fixed threads described a-bove, we usually find also a certain propor-tion of free threads. These are either car-ried passively in the water or they possess an independent power of movement. Some of them may reach as much as one milli-metre in length. The free threads differ from the attached ones in that the sheath in them does not appear to harden, but persists as a soft clinging mantle throughout the whole of the free life of an individual.

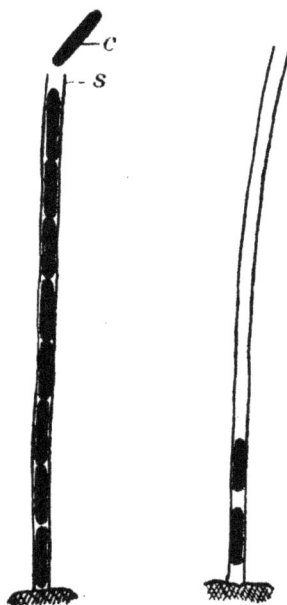

FIG. 38.

FIG. 39.

FIG. 38.—Cladothrix dichotoma. Thread out of which the developing cells are escaping successively from the apical opening of the sheath.

c = escaping cell.

s = sheath (× 1000).

FIG. 39.—Cladothrix dichotoma. Old thread composed of sheath which still contains two of the cells with which it was originally filled (× 1000).

The Transverse Septa of the Cladothrix Sheath.—
The appearance and structure of these septa has
already been described. Hitherto no mention of
them has been made by previous writers. It has
already been stated that when a cell slips out of the
sheath laterally, the space originally occupied by the
cell becomes filled by an extension into it of sheath
material. When this happens the continuity of the
cells is interrupted by a broad bridge composed of
the same mucinous material as the sheath (Fig. 37).

Büsgen (1) has noticed these wide bridges but has
come to a wrong conclusion with regard to their
significance. On page 151 he says : "Von an-
deren Entwickelungs-ständen der Cladothrix traten
in meinen Culturen nur noch eigentümlich gestaltete
Faden auf, welche sich durch besonders hormogonien-
bildenden Algen vergleichen lassen. Dieselben
beziechneten sich durch besonders dicke Scheiden
und eine auffallende Gruppirung ihre Stäbchen aus.
Oft ebenso lang wie breit, lagen die letzteren bei-
spielsweise in weniggliederigen Reihen zusammen,
welche durch Pfropfen einer structorlosen Masse,
wohl Reste abgestorbener Stäbchen, getrennt waren."

If the illustration mentioned by him be compared
with Fig. 37, it will be seen that his subdivision
of threads into sections bounded by "hormogonia"[1]

---

[1] In the filamentous Cyanophyceæ vegetative propagation is
effected by the breaking up of the filament into lengths, each such
portion being termed a *hormogonium :* in most of them the limits
of the hormogonium are indicated by large inert cells called
heterocysts.

similar to those that appear in the Cyanophyceæ, was founded on a misconception as to the nature of the broad bands which appear to divide each thread into a number of well-defined zones separated by "hormogonia". The limits of each of his "hormogonia" were therefore determined by the length of the thread intervening between two points at which cells had been released laterally. To denominate the strip of cells lying between one wide bridge and the next a hormogonium is to miss entirely the significance of this term, and doubtless the name would not have been applied by Büsgen had he known that between *every* cell in the "hormogonium" there is a septum of the same nature as these wide bridges, but of much narrower width (Fig. 36).

Absence of Special Organ of Attachment.—I failed to discover any special organ of attachment, the adhesiveness of the mucilaginous sheath being apparently sufficient to cause the adhesion of a thread to any particulate matter in the water. It is probable that under certain circumstances when the need arose a greater amount of sheath material would be formed at the base to secure better attachment of the thread to its support. I have, however, never observed any difference in this respect between the base and any other part of the thread. In Fig. 45 is shown a colony coherent to a small lump of mucilage, and the individual filaments of a colony are usually held together in this way. The mucilage can scarcely, however, be designated an essential part of the organism. It is obviously of a temporary nature, for this formation of colonies is not a normal condition for Cladothrix. Büsgen speaks of an adhesive disc in connection with Cladothrix. This was probably caused by an accumulation at the base of sufficient of the mucilaginous sheath material to create the appearance of a special organ for the purpose.

RESERVE MATERIAL. MEMBRANE OF THE CELLS

In young cells *oil-bodies* are commonly found. These have already been noticed by Büsgen (1). They are usually absent from the cells of filaments taken from artificial cultures. This is not surprising when we consider that in such cultures Cladothrix does not obtain food of equal value to that which it obtains in a state of nature ; in fact, in artificial cultures the threads are probably in a half-starved condition. In many cells I have found *glycogen* as a reserve material either alone or in addition to oil globules (Fig. 40).

By appropriate staining each cell is found to be limited externally by a well-defined membrane. This structure stains a deep brown when iodine is added, this being the same colour as that assumed by glycogen when treated with this reagent. As the removal of glycogen from the cell proceeds hand in hand with a diminution in the capacity of the membrane to react to iodine, it is probable that this diminution of staining capacity is caused by the *removal of glycogen from the membrane.* It must be remembered in support of this statement that the membrane of bacteria is not, like the cell membranes of the higher plants, a secretion of dead matter, but is to be regarded as being rather the outermost and densest layer of the cell, and is in all probability not devoid of living matter. It is, however, very sharply marked off from the rest of the cell. It is interesting to find the substance of the cell membrane utilised for storage purposes. The glycogen is evidently the source of the carbohydrate supply, as it so often is in the lower fungi, and takes the place of the starch reserves of the higher plants.

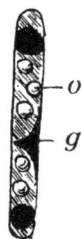

FIG. 40.—Cladothrix dichotoma. Single cell containing glycogen (*g*) and oil globules (*o*) (× 2000).

The combination of glycogen and oil is often encountered in the lower bacteria, as, for example, in Bac. tumescens. When Cladothrix is removed from its natural habitat and put with water into a beaker, the cells grow and multiply for a short time, living on the food supply contained in their cells. If the cells be examined twenty-four hours after being placed in the beaker, the amount of glycogen and oil will be found to have diminished considerably, and after 48 hours will, in most cases, have disappeared altogether. With regard to the nitrogenous supply of food we can only conjecture that it is present either in a state of solution or in the form of very minute granules scattered throughout the body of the cell.

Cell-division.—The method of division is in all

6

essentials the same as that which prevails in Crenothrix and in the Bacteriaceæ. This consists in the
development of a transverse membrane across the
cell. First of all the cell elongates slightly and then
forms the transverse membrane. Following this a
constriction develops at the dividing area, giving the
cell a distinct waist. From this point on the division
of the cell is the result of three processes which take
place concurrently—

    a. The elongation of the dividing cell.

    b. The thickening of the transverse membrane.

    c. The deepening of the constriction.

The separation of the daughter cells follows as
a result of these activities. They move apart and
repeat the process if conditions are favourable. Cell-
division at first takes place uniformly among all the
members of a thread, but after the hardening of the
sheath it is probably confined to the more basal cells.

Reproduction.—There is essentially only one type
of reproduction, namely, by the rejuvenescence of a
single cell or a group of cells. Each cell is thus
potentially immortal. There is no trace of the formation of spores of an asexual or a sexual nature. It
must be borne in mind that the different methods of
reproduction described below differ only in degree
and not in kind.

    1. *By Liberation of Thread-fragments.* — This
method was first observed by Zopf in the case of
attached filaments. Portions are cut off, these move
away under their own power and in turn elongate to

form new filaments. I have observed this liberation of thread-fragments but with this difference, that
the released portions swam off not as straight but
as *spiral* threads. The nature of these spiral threads
will be dealt with separately in a later section. Some
of the straight thread fragments observed by Zopf
attained a length of 1-$1\frac{1}{2}$ millimetres. Each thread-
fragment is made up of a varying number of cells
wrapped, as usual, in a thin covering of invisible
mucilaginous matter which later develops into the
sheath.

2. *By the Liberation of Rejuvenated Single Cells.*—
This mode is by far the commonest in the life history
of this organism. As explained above, the pressure
exerted by the growing cells serves to prevent the
sheath from hardening so as to leave no outlet for the
cells to the outside. This pressure serves to keep
open the apex of the sheath, and after the hardening
is complete the opening naturally becomes a permanent one. Out of this vent is thrown the topmost
cell at the apex, and its place is taken by the next,
which in turn suffers ejection (Fig. 38). In some
cases the cells are liberated merely by pressure from
behind. In other cases, however, they swim out,
motility being accomplished by the lashing of the cilia
with which they are beset. In Fischer's continental
variety of Cladothrix the ciliation was sub-polar (Fig.
41). In the British variety I found the ciliation to be
exactly polar (Fig. 42). Hoeflich refers to these
liberated cells as "spores". It seems clear that there

is no feature in which they differ from any of the other cells inside the sheath. It is true they develop cilia but so do any or all of the cells when circumstances are favourable. There is no ground for regarding these liberated motile cells as other than vegetative cells.

3. *The Formation of Spiral Threads.*—In 1882 Zopf (1) announced that in certain circumstances spirally shaped threads are liberated from filaments of Cladothrix dichotoma, and that these fragments behave in all respects like Spirilla. Since his announcement adverse criticism has appeared from *Winogradsky* (1), *Büsgen* (1), and *Hoeflich* (1), all of whom have professed themselves very sceptical of the accuracy of Zopf's observations. By keeping artificial and natural growths of Cladothrix under observation for about 18 months, the present writer (Ellis (7)) was able to confirm and extend Zopf's interesting discovery. The phenomenon occurred during this period in one culture only out of dozens under examination. This was a seven-day old culture in an open beaker. The medium of cultivation was water to which a small quantity of ferrous carbonate and a very minute quantity of peptone ('05 per cent) had been added. Between the 7th and the 10th day the threads of the culture broke up into various sized fragments. On the 7th day the tufts of Clado-

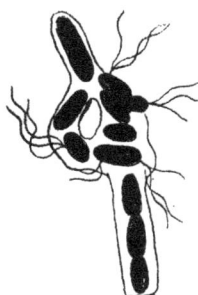

FIG. 41. — Cladothrix dichotoma. Motile vegetative cells escaping from apex of thread. Cilia in subpolar position. (After Fischer.)

thrix filaments were observed to be in a state of violent agitation. The surrounding water contained Spirilla that, in the unstained condition, could not be distinguished from typical members of the Spirillaceæ. Cilia-preparations showed *spiral filaments* of Cladothrix in the water, and it was evident that the violent wriggling of these spiral forms in the operation of freeing themselves from the filaments had caused the agitated convulsive movements of the filaments. The cilia preparations *showed spiral filaments in all stages of liberation.* In Fig. 42 one is shown which has been newly liberated. The cilia in all cases were polar. In some (Fig. 43), where the spiral thread consisted of 4 or 5 cells, each cell had a single cilium at the ends. Most of the spiral threads, however, possessed a small number of cilia at each end of the thread (Fig. 44). Writing of these spiral threads Winogradsky (1) (p. 352) opines that their ultimate fate does not settle the question of their origin. In the culture mentioned above I was able to observe the liberation of spiral fragments in all stages. Each fragment had not only an undulatory form and movement but possessed a sinuous form and polar cilia (Ellis (7)). In Fig. 42, at "*a*" is shown the place occupied by the fragment—here composed of only one cell—which lies just outside the sheath from which it has escaped laterally. The evidence is complete as to the origin of these spiral fragments from the Cladothrix threads. The spiral form is evidently correlated with the possession of strong polar cilia and a very tenuous membrane, and the assumption of this form is the logical sequence of the possession of such attributes. We need not claim kindred for this organism with the genus Spirillum on account of the possession by both of an undulatory movement

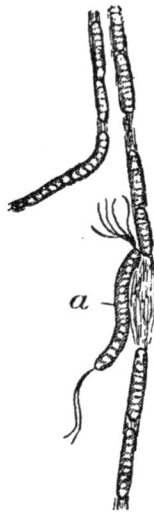

FIG. 42.   FIG. 43.

FIG. 42.—Cladothrix dichotoma. Shows escape laterally of one of the cells. This cell (a) has developed polar cilia and assumed a spiral form. Stained with anilin-gentian-violet (× 800).

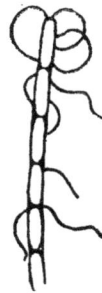

FIG. 43.— Cladothrix dichotoma. Fragment liberated from thread, one or two cilia at poles of each cell. Stained with anilin-gentian-violet (× 800).

or a sinuousness of form, but the fact of such a possession by both does suggest that the distinction between a Bacillus and a Spirillum is not as great as is commonly believed. As to the ultimate fate of the spiral fragments of Cladothrix, why should it differ from that of the straight fragments? There can be little doubt that they settle down as do the straight fragments, and elongate to form fresh threads if the conditions are favourable. It is very probable that in course of time the cilia drop off, and the sheath gradually hardens, with the result that the straight form is once more assumed by the Cladothrix of the next generation. There was nothing in the phenomenon to suggest that the assumption of the spiral form was the prelude to the adoption of pleiomorphic changes ; that, for example, the liberated spiral fragments assumed a mode of life characteristic of the genus Spirillum. We see here merely another phase or another modification of the vegetative mode of reproduction which alone prevails in Cladothrix dichotoma. The reproduction of this organism may be summarised by stating that it consists of the liberation of either single cells, or of thread fragments made up of cells. In the former case the cells may be straight or curved, motile or non-motile, and, if motile, the cilia may be placed in a polar or sub-polar position. Finally, the expulsion may be effected from the top or through the sides. In the latter case the thread fragment may be composed of a few or of many cells, may be motile or non-motile, and may be straight or thrown into undulations.

MOTILITY AND ORGANS OF MOVEMENT

Both Hoeflich (1) and Fischer (1) have examined the continental variety of Cladothrix for cilia, and both have ascertained that when any individual cell of this species was motile, the motility was due to a small number of cilia bunched together in a sub-polar position. In the British variety the ciliation was invariably polar. Hitherto no investigations on the ciliation of the motile *filaments*—each of which is, of course, made up of a varying number of cells—have been conducted. Such filaments, as explained above, are usually straight, although very occasionally, as shown above, they adopt the spiral form. In both cases the cells when motile possess cilia only at the poles. In the spiral filaments the cilia are found only at the extremities (Fig. 44). In the straight filaments I have found examples with cilia only at the ends, and others in which they were present, one or two at the pole of each cell, or the majority of the cells, in a filament (Fig. 43); but in no single case was the arrangement of the cilia in the British representatives sub-polar. The variation in the mode of arrangement of these structures in a species of world-wide distribution is interesting, and we may recall the fact that among the lower bacteria those which effect the oxidation of ammonium compounds into nitrites, differ throughout the world in little other than a slight difference in the form and insertion of the cilia.

FIG. 44.—Clado-thrix dicho-toma. Fragment liberated from thread. It consists of 5 cells inside a sheath, possesses cilia only at the ends and is spiral in form. Stained with anilin-gentian-violet (x 2000).

MORPHOLOGICAL AND PHYSIOLOGICAL VARIATIONS

In the slimy particles of dirt which are found in sewage and in the waste waters of factories there is sometimes found an organism that closely resembles

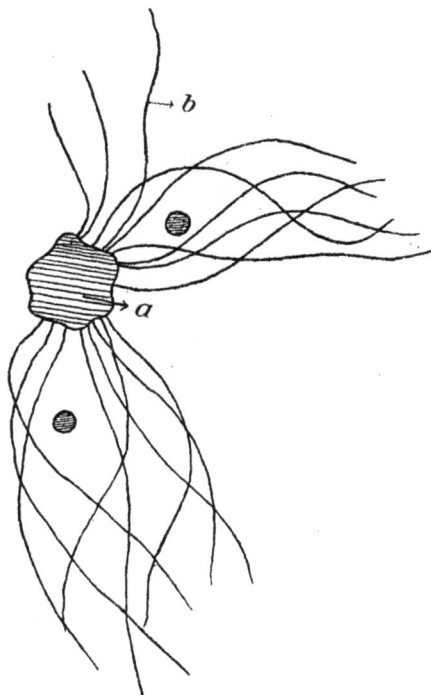

FIG. 45.—Cladothrix dichotoma.   Diagrammatic representation of appearance under the microscope of one of the grey-white flecks which appear in an artificial culture of Cladothrix dichotoma.   Each one of these flecks is made up of a number of Cladothrix threads (*b*) attached to a central mucilaginous mass (*a*).

Cladothrix dichotoma, but differs from it in that the threads usually run parallel, all being enveloped by a common mucilaginous covering.   This organism is named Sphærotilus natans.   Unfortunately it has been very little studied, and it seems not unlikely that the peculiarity in the nature of its mucilaginous covering is associated with its environment, and is purely local and of a temporary character, that, in short, Sphærotilus natans is a variant of Cladothrix dichotoma. Migula (1) has described Cladothrix dichotoma as a "sammel-species," and there is every reason to believe that further investigation will show that Cladothrix consists of a group of variants clustering round one or two dominant types. Another morphological variety is possibly to be found in Clonothrix fusca (Schorler), with which, as one of the iron-bacteria, we shall deal with in a subsequent page. This differs in possessing a sheath which tapers to a point at the apex, but is otherwise similar in essential respects to Cladothrix dichotoma. Whether we must regard a type of Cladothrix with "false dichotomy" as a third variety, it is difficult to say, but it is conceivable that a slight alteration in the constitution of the sheath might impart to that organ greater adhesive properties, sufficient at any rate to exercise a retarding influence on the laterally escaping Cladothrix cells. Prevented from escaping, the cell that had won to the outside would then elongate to form a filament which would have the appearance of a branch arising from the parent thread. In this way "branches" would arise from different parts of the thread, and as the "branches" themselves would subsequently form others of a similar nature, it can readily be seen how a tree-like colony of threads would arise superficially, quite unlike the dominant type which we have discussed above. If we regard the Cladothrix showing this false dichotomy as also a variant we shall have three morphological varieties centring round a dominant type. These three are—

1. Sphærotilus natans.
2. Clonothrix fusca.
3. Cladothrix dichotoma with false dichotomy.

It will be noticed that all three differ from the dominant Cladothrix dichotoma, in showing a variation in the structure of the sheath. Coming now to the cells, we find that the British type described above differs from the type examined by Fischer (1) and by Hoeflich (1), in the possession of polar in place of sub-polar ciliation. Possibly other variations in structure also exist, and it is probable that extended research will reveal their existence. So far as the investigations have already proceeded they show us a simple type of organism with a simple vegetative mode of reproduction. The possibility of attaining a more complex and elaborate structure seems to have been achieved by the sheath, rendering possible the cohesion of a number of similarly constituted cells. As a next step we may expect a differentiation in the functions of the cells, and Cladothrix and its attendant satellites seem to be at a stage when a commencement of differentiation has been made. At present, however, we are unfortunately not in a position to state whether or not the peculiarities which distinguish the variants from the presumed parent stock reproduce their peculiarities in their offspring.

With regard to the name by which this group should be designated, it seems advisable to retain the name Cladothrix dichotoma for both the British and the continental forms, with the mental reservation that later researches may make a finer discrimination necessary. In the case of Sphærotilus natans and Clonothrix fusca, although suspect, they have not yet been proved to be variants of Cladothrix dichotoma, so it would not at present be wise to change even their generic names. It is certain at any rate that if they are distinct species, their alliance to Cladothrix dichotoma is of a very close nature. Assuming that they are specifically but not generically distinct, of the three claimants to the generic name, viz. Sphærotilus, Cladothrix and Clonothrix, the second has become so fixed and the organism bearing that name is so much more universally known than the other two that it would be advisable to adopt the term Cladothrix for the generic name, in spite of the priority in time possessed by the term Sphærotilus. Again, while Sphærotilus is an obscure organism, Cladothrix is the central figure of a collection of species of which Sphærotilus is probably a member. The term Cladothrix may, therefore, be said to have earned, as it were, the right to denominate the group. Before closing a consideration of the morphological varieties, we must mention *Chlamydothrix sideroporous*, which is described in a subsequent section. This organism is obviously a variety of Cladothrix dichotoma, from which it differs in no respect except that it possesses an adhesive disc. We have already shown the facility with which the sheath fills with its material any unoccupied cell-compartments (Fig. 37), and it can scarcely be doubted that the formation of an adhesive disc is well within its powers. In the meantime this organism is best regarded as another variety of Cladothrix dichotoma, until it can be shown that the disc is not an organ of purely local significance called into being by the play of factors of only temporary importance. The investigations on Cladothrix have already revealed the fact, as attested by Migula, that while some varieties of Cladothrix attract iron compounds, others are indifferent to this substance. Extended investigation will probably add to this list, but so far this field of research has scarcely been trodden.

PHYLOGENETIC RELATIONSHIPS

Apart from Sphærotilus natans and Clonothrix fusca, the generic and even specific distinctness of which are doubtful, Crenothrix polyspora is the organism most closely related to Cladothrix. The type remains the same in both, namely, a series of vegetative cells enclosed in a mucinous covering, which later hardens to form a tubular, more or less rigid sheath. There is no trace in either of reproduction, other than the purely vegetative mode which consists in the liberation and continued development of vegetative cells. While, however, in Cladothrix there is throughout life only one row of cells in a filament, in Crenothrix this arrangement persists only

in the younger stages.   In older filaments, as described above, division takes place in three directions
of space, with the consequent formation of a cluster
of vegetative cells inside the sheath, and the expansion of the latter at the apex to form a trumpet-shaped
structure.   Added to the fact of their close intimacy
in ferruginous waters, and the absence of colouring
matter from either, there can be little doubt either
that both have sprung from a common stock at no
very remote period, and that Crenothrix has advanced a little further in the march of evolution, or
that Crenothrix is directly descended from Cladothrix.

We shall consider other relationships of this group
after we have described the other iron-bacteria.

CHAPTER VI

CLONOTHRIX FUSCA (*Schorler*)

THIS organism was discovered by Schorler (1) in the water-works of Dresden and Meissen in Germany, and although it has not subsequently been found in any other place, it is conjectured by its discoverer to possess a fairly wide distribution. Young threads are 2μ to 3μ thick, older ones 5μ to 7μ. When the latter become covered with manganese or manganese and iron they may reach the enormous thickness of 24μ. Schorler (1) states that false branching similar to that observed in Cladothrix occurs. At the point of origin of one of these "branches" a club-shaped swelling usually arises, formed by an increase in the output of mucilaginous matter. The cells are cylindrical, some being long, others short : they measure approximately 2μ in thickness. Further, the cells are visible without special treatment, in this respect resembling Crenothrix rather than Cladothrix. As is the case with these two organisms, the cells of Clonothrix slip out either laterally or apically. According to Rullmann—for I have not had access to the original paper—Clonothrix differs from Clado- thrix in that the "gonidia," as he calls the liberated cells, are formed by the repeated division of some of the short cells inside the sheath, in this respect, of course, resembling the " microgonidia " of Crenothrix polyspora. The three most striking features to which Schorler calls attention as those which establish a generic distinctness to the organism are :—

1. The production of " gonidia " by the splitting of short cells inside the filament into a number of short rounded cells. These cells are the gonidia which form new filaments.

2. An extraordinary capacity for storing iron and manganese on the sheath, the thickness of the encrusted sheath being sometimes twelve times as much as that of the original filament.

3. The formation of very short disc-like cells inside the sheath.

It has still to be proved that these features are constant. The second and third of the three characteristics may be due to local conditions of environment. The first, however, appears to be a feature distinctive of Clonothrix, and to bestow a generic distinctness to this organism. This new member of the group is of great interest for it seems to mark a new road in the production of new varieties from the well-established Cladothrix dichotoma to which it is obviously as closely related as is Crenothrix polyspora. It is to be hoped that further investigations will be forthcoming which will throw additional light on its relationships.

OTHER IRON-BACTERIA

1. In 1909 a new and interesting form of iron-bacteria was discovered by Molisch.

Chlamydothrix sideropous[1] consists of an attached thread which may reach a length of 600μ or more, and a thickness of 0·6μ. The disc has a diameter of 6-30μ. The thread is colourless, unbranched, and by appropriate staining it can be demonstrated that it consists of a sheath enclosing a number of cells. As the sheath gets older it gradually becomes brownish-red by deposition of ferric hydroxide. We have already given our reasons for the opinion that this organism is in all probability a variety of Cladothrix dichotoma, which under the influence of local conditions has developed an adhesive disc. It is not probable that the power to develop this organ is inherited in some varieties of Cladothrix and not in others.

2. Molisch has also described two organisms which belong to quite a different category to any of those that we have already described.

In Bacteriology we are now getting accustomed to find the number and kind of organisms possessing certain physiological properties becoming much greater

---

[1] This generic name was introduced by Migula, with the following characteristics : cells non-motile, cylindrical, unbranched ; they are further surrounded by a *sheath* forming threads which show no distinction between base and apex. To this genus he transferred Leptothrix ochracea and Gallionella ferruginea.

7

in number and different in kind, on extended investigations being made.   The history of the iron-bacteria
is no exception.   At first it was considered that the
power of promoting the oxidation of ferrous to ferric
compounds was confined among the bacteria to a
small class of organisms, all more or less closely
related.   It had become customary, either con-
sciously or unconsciously, to associate the iron-
bacteria with the thread-bacteria.

In 1909, however, a very peculiar organism was
described by Molisch (2), which lived as an epiphyte
on water plants.   Its presence is indicated by the
appearance of a rust-coloured crust on the surfaces of
these plants.   In the crust itself rounded clear spaces
occur, each about 1.8μ to 3.6μ in diameter.   In the
spaces small round cells or cocci are found, and it is to
the activity of these minute cells that the formation of
the crust is due.   Each clear space encloses a colony
of cocci, and the crust of iron invariably begins round
these colonies.   This organism was named *Siderocapsa Treubii* by its discoverer.   A second species of
the same kind, found by the same investigator, has
been named *Siderocapsa major*.

In 1913 an organism was described by Mumford
(1) under the title "A New Iron-bacterium".   This
member belongs to the rod-shaped bacteria (Bacteriaceæ).   As will be shown in later pages, the
evidence in favour of "Bacillus M7," as the organism
was named, being a member of the iron-bacteria, is
very inconclusive.

These and other researches prove that it is highly probable that many of the lower bacteria, as well as the thread-bacteria, function permanently or temporarily as iron-bacteria. Investigations on the physiology of these bacteria have completely altered the views that were formerly held on the nature of the activities of these organisms.

These questions are described in Chap. VIII. We may, however, here again anticipate by stating that the chief activity of the iron-bacteria lies in an absorption of organic matter in solution, and that it is the absorption and oxidation of this organic matter that is primarily responsible for the liberation out of solution of the iron which was bound to the organic radicle. That many of the most common of the lower bacteria will be found to be iron-bacteria in this sense of the term seems almost a certainty. We may, however, dismiss the idea that the iron-bacteria obtain their energy by the oxidation of ferruginous compounds. This idea originated with Winogradsky, and was based on an analogy with the sulphur-bacteria, which do capture the energy liberated when sulphides are oxidised first to sulphur and then to sulphates. It never went further than the argument from analogy, and no proof has yet been forthcoming. We shall show, however, that the iron-bacteria do influence the rate of this oxidation in the water in which they grow, not by capturing the energy, but by liberating substances which affect the oxidation as an external process. It is hardly likely that any iron-bacteria will be found in which the oxidation is a physiological process under the direct control, as it were, of the protoplasmic molecule.

Mumford has claimed to have found an "iron-bacterium" in the Bridgewater Canal tunnels at Worsley, Lancashire. A bacillus was isolated and named, provisionally, "Bacillus M7". He describes its growth in various media, and he also made some physiological tests. He assumed that it was necessary to supply this organism with some ferruginous compound, and added ferrous ammonium sulphate, and ferric ammonium citrate to his culture solutions. He records the complete precipitation of these substances under such circumstances, and assumed that this precipitation was due to Bacillus M7. Unfortunately, the author has laid himself open to a *non sequitur* fallacy, for in no case is the *causal* connection between the incidence of the growth of the microbe and the oxidation of the iron compound made clear. The growth of the organism was guaranteed by the addition of peptone, and it is unfortunate that no control cultures were instituted to prove the necessity of the iron compound to the growth of the microbe. Again, no comparison was made between the rate of precipitation of the ferruginous compounds in a liquid medium in which the microbe was under cultivation, and the rate in a culture in which the same compounds were present, but in a sterilised condition. Under such circumstances we should have been able to determine the exact part played by Bacillus M7 in the process of oxidation. One suspects that, in the medium in question, Bacillus M7 would have grown quite as well without as with the addition of the iron. Mumford also prepared an enzyme from the products of the activity of the bacillus, and claimed to have effected the oxidation of ferrous compounds through the agency of the enzyme. Unfortunately, again we miss the control. He should have instituted a comparison of the effect on the ferruginous compound of the liquid containing the enzyme, and the same liquid rendered ineffective by the destruction of the enzyme by heat or any other destructive agency. The investigation appears to the present writer to have proved neither that the organism exerted any direct influence on the precipitation of ferric hydroxide, nor that the presence of iron was in any way a necessary condition for the growth of the organism. At the same time, the field of research opened up by attempts in this direction is a valuable one, and it is to be hoped that further investigations will be prosecuted.

Some years ago the present writer published a preliminary note on the discovery of five new species of iron-bacteria (Ellis 4)). These were named—

      1. Spirosoma ferrugineum.
      2. Nodofolium ferrugineum.
      3. Leptothrix Meyeri.
      4. Spirophyllum tenue.
      5. Spirosoma solenoide.

On further investigation I found that Spirosoma ferrugineum was a sinuous variety of Leptothrix ochracea devoid of specific independence, for it was found possible to trace stages of transition from the straight form of Leptothrix ochracea to the sinuous variety of Spirosoma ferrugineum. I am now convinced that Nodofolium ferrugineum is an extreme form of Spirophyllum ferrugineum, one in which the spiral twists which are noticeable in the latter organism have been carried much further. If we imagine the twisting of the Spirophyllum band carried on until several complete twists have been taken, we can see how the organism will depart from the loosely twisted shape and assume a more definitely segmented form. The "loop" in such a figure is caused by the broad side of the band being presented to the eye, while the "node" is caused by the presentation of the band to the eye in the "end on" position. If a band of elastic be subjected to two or three complete twists, it is obvious that some parts of the band will be seen broadside-on, while other parts will be end-on to the eye. With regard to Leptothrix Meyeri, I have fortunately been able to prove that this structure described as a modification of Crenothrix by Schorler, denominated Leptothrix Meyeri by myself (Ellis (4)) and Megalothrix discophora by Schwers (1), is nothing other than the basal part of Crenothrix deprived of its cells and in a condition of mucilaginous degeneration (see Chap. IV), (see Plate IV, Fig. 3).

I have not since 1908 encountered either Spirophyllum tenue or Spirosoma solenoide, and cannot state whether they form well-defined organisms with a specific distinctness or whether they are part of pleiomorphic changes in the life history of one of the known iron-bacteria.

Affinities of the Iron-Bacteria

When we refer to the Dicotyledons as those plants the seeds of which are characterised by the possession of two cotyledons, we know that the use of this characteristic for purposes of division will bring into one group a number of plants that have many other properties in common, in addition to the possession of two cotyledons. The division is a *natural* one. When, however, a number of organisms are placed in one group on the strength of the possession by them of a certain physiological peculiarity in relation to iron compounds, we cannot *expect* to find that these organisms possess any close affinities. The division is an *artificial* one. It neglects the affinities imposed by nature, and introduces a basis of division almost as artificial as that adopted by librarians which bands together the books of those authors whose names begin with the same letter. When, therefore, we consider the affinities of the iron-bacteria, we must not expect to find a group of closely connected bacteria, and it seems very likely that in a few years we shall have representatives of iron-bacteria from every one of the known classes of bacteria. However, we

do find within the iron-bacteria that certain species
cling together, as it were, more closely than do others,
forming a *natural* sub-group.   As is to be expected
the members of the sub-group are connected not only
with species within the iron-bacteria but also with
others that are outside this group.   Crenothrix polyspora, Cladothrix dichotoma, and Clonothrix fusca
form one closely related group, and have probably
sprung from the same common ancestor.   Their
relationships may be expressed as follows :—

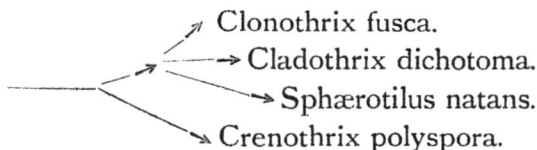

```
                    Clonothrix fusca.
                ———→ Cladothrix dichotoma.
                ———→ Sphærotilus natans.
                ———→ Crenothrix polyspora.
```

As already stated it is not certain that Clonothrix
fusca and Sphærotilus natans are to be regarded as
specifically distinct from Cladothrix dichotoma, but if
so their separation from Cladothrix or an ancestor
of this organism must be regarded as having been
effected within comparatively recent times.   Cladothrix dichotoma is at present the most widely distributed, the most highly variable, and possessing the
greatest number of individuals among these three ;
it seems therefore certain that it is the centre of an
activity out of which new forms are arising with
specific distinctness.

The life histories of Leptothrix ochracea, Gallionella
ferruginea, and Spirophyllum ferrugineum are so
similar, and their association so close, that not only
do they form a very closely allied group but it is

questionable whether they are not all three manifestations of one and the same organism.   This group is
not closely related to the one that we have just described.   It is too premature yet to discuss the affinities of the organisms other than the above, which
have been described as iron-bacteria.   If they are
capable of expediting the oxidation of ferruginous
compounds we can regard them only as members of
the lower bacteria which have gradually accustomed
themselves to the exigencies of their new ferruginous
environment, and not as organisms genetically connected with the species already included under the
iron-bacteria.

   If we seek for affinities to the iron-bacteria among
other organisms, it is clear that these will depend on
the group of iron-bacteria which is under consideration.
Cladothrix dichotoma marks a simple step from the
genus bacillus.   If we imagine the dividing cells of a
bacillus remaining coherent and surrounding themselves with a sheath, the form of Cladothrix would
be reproduced.   In spite, however, of its more complicated vegetative structure, the reproduction of
Cladothrix is of a more primitive type than that
which prevails in many of the bacteria, inasmuch as
in the latter asexual reproduction by the formation of
endospores is well marked, while in Cladothrix, in
spite of its greater complexity of form, the reproduction remains purely vegetative.   We may conjecture
that both Cladothrix and the spore-forming bacteria
have descended from a more primitive bacillus-like

ancestor, which developed along two lines, one in the direction of a more efficient method of reproduction, the other in the direction of a more complicated vegetative form, resulting in the evolution of Cladothrix. From this organism would follow Crenothrix, and the satellites that group round Cladothrix. Whilst the connection of this group with the lower bacteria is so well marked, the same cannot be said of the group that cluster round Leptothrix. Here we appear to be dealing with fungal forms that, owing to their peculiar habitat, have changed their filamentous form, and have assumed one that conforms more readily to the conditions that prevail in ferruginous streams. The method of reproduction by the budding-out of conidia from the surface is a characteristic quite alien to bacteria and to the other iron-bacteria. In the same way we are precluded from seeking affinities among the algæ, many of which resemble superficially Leptothrix or Gallionella or Spirophyllum. Among the fungi, however, conidia-formation or budding is a familiar phenomenon. It is probable that we must look for the nearest relatives of the Leptothrix group among the lowly representatives of some of the great groups of the fungi. Thus the yeast plant among the Ascomycetes forms essentially the same kind of reproductive cells by a process of budding, and does, further, occasionally assume a thread-like form. We could in this way establish contact with several of the primitive or degenerate representatives of several of the great fungal groups.

A general survey of the life histories of these three organisms inclines one to attach them to the lower fungi rather than to the other bacterial orders. The fungal origin is emphasised by the knowledge that the whole of the iron-bacteria are genuine saprophytes dependent on the presence of organic matter for the maintenance of the food supply. It is hopeless to seek for closer affinities among any particular group of fungi but it would not be inappropriate to place Leptothrix and its close allies in a place reserved for the lower members of the Phycomycetes.

The departure from the form of Leptothrix to that of Gallionella is a slight one, for both are essentially thread-like in form. The only difference is that Gallionella has developed a sensitiveness resulting in a change of form when brought into contact with other objects or when one part touches another part, while in Leptothrix the development of a stiff membrane has precluded any but the straight forms being evolved. The change into the flattened shape assumed by Spirophyllum is more difficult to understand, but however great the change in external form, it cannot weigh against the close similarity at all points in their life histories, which exist between Spirophyllum and Leptothrix and Gallionella. This similarity is so great that it precludes the supposition that a closer alliance may exist between Spirophyllum and such algæ as Spirulina which it resembles in general appearance.

CHAPTER VII

THE ARTIFICIAL CULTIVATION OF THE
IRON-BACTERIA

TWO kinds of artificial cultures must be distinguished: 1. Pure Cultures. 2. Impure Cultures. In the first case the organism under investigation is the only one present in the culture medium. Obviously it is advisable if possible to procure this kind of culture, and for the investigation of physiological problems it is almost indispensable. An impure culture, on the other hand, contains, besides the organism under investigation, others which it has been found impossible to eliminate. For the investigation of the details of the life history of any particular organism an impure culture is sometimes of greater value than the other kind, as the growth of the organism is observed under the conditions which persist in nature. The advantage holds only if the organism under investigation is clearly marked off from the others in such a way that there is no risk of confusing one with the other. To the experienced observer the identification of the different species of iron-bacteria is not difficult, and the impure artificial culture can therefore be relied upon to give satisfactory results.   The usual method of procedure is to remove samples of iron-water from their natural source, in sterilised flasks, and then to add various substances in order to watch the effect on the various organisms in the iron-water.   If growth is stimulated in the iron-bacteria, the developmental phases are carefully noted and described.   As will be shown below, pure cultures have been obtained in regard to some of the iron-bacteria, and the methods by which these were obtained will be noticed in their proper place.

It must be borne in mind that the development of organisms does not necessarily or even usually follow the same course in pure artificial cultures as in nature. To take a well-known example, the long branching threads that Bac. tuberculosis often develops in the body are never encountered in test-tube pure artificial cultures.   Organisms must not be regarded as chemical reactions which take a stereotyped course under all conditions ; they adapt themselves to suit the different environments in which they grow, and the adaptations are comparatively great in the case of such plastic organisms as the bacteria.   One must therefore enter a caveat against the attitude adopted by Molisch who finds himself unable to believe in the existence of reproductive structures which are not forthcoming in his own artificial cultures.

We may now consider in detail the results that have followed attempts at the artificial cultivation of the iron-bacteria.

Crenothrix Polyspora.—Several successful attempts

at the artificial culture of this organism have been
made. The first was accomplished by Rössler (1) in
1893. This writer states that for two years he
cultivated Crenothrix polyspora with the best results.
He had found that a part of a 25 centimetre thick
bricked canal-wall had within three years become
eaten through and through with this organism. For
two years he appears to have cultivated the species on
these bricks. He thus obtained and carried on impure artificial cultivations. The liquid medium was
water—presumably the canal water—to which ferrous
sulphate had been added. It is probable, nay, almost
certain, that the addition of ferrous sulphate was
superfluous, the organic constituents in the canal
water being sufficient. Rössler made his cultivations
at a time when the presence of a ferruginous compound
was regarded as a *sine qua non* for the growth of
iron-bacteria, and he does not seem to have attempted
to induce growth in the absence of the iron compound.
It is evident that the canal water contained all the
ingredients, organic and inorganic, that were necessary
for growth, and he seems to have repeated on a small
scale what was done on a large scale in the canal
on its bricked walls. The artificial cultivations of
Garrett (1) are not so easily explained. This writer
informs us that he made attempts to cultivate the
" cocci " during the 1896 calamity and that they were
successful. He obtained growth on a gelatine plate,
on agar-agar jelly at 37° C., and on potato. On the
last-named a buff-coloured, somewhat crumpled ex-

pansion covered the surface in twenty-four hours. On coagulated serum at 37° C. in one to two days an almost colourless tough expansion was obtained. Unfortunately Garrett does not state the results of the *microscopic* examination of the growths in his culture media, and he seems to have assumed that all the growths were due to Crenothrix polyspora. His inoculation-material doubtless contained a plentiful supply of Crenothrix, but there may, in fact, there must have been other bacteria as well, and the growth on the culture media may have been caused by a development of these bacteria. Under abnormal circumstances it is true that Crenothrix is capable of what may be called an extravagance of growth, but it may safely be asserted that in normal times Crenothrix will not grow in the media used by Garrett, and as in this case the inoculation-material was impure, we must reserve our judgment on the nature of Garrett's growth on *solid* media. There is not the same objection to his *liquid* culture media. Some of the reservoir water was placed in a beaker and inoculated with a small portion of the red fluid drained from the sand at the bottom. In this medium the cocci multiplied in Zooglœae and ultimately formed a deposit at the bottom of the beaker. From the cocci filaments subsequently formed. During this process the water changed its colour from red to green, and within a few hours a yellow-green deposit was formed at the bottom of the tube. The multiplication of Crenothrix polyspora, under abnormal circumstances, in the form of cocci enclosed in the mucilaginous masses known as Zooglœae is well known. Presumably the redness referred to by Garrett was due to the formation of the reddish-brown ferric hydroxide. Most waters of this kind become green on exposure to the sun owing to a preponderating growth of one or other of the lower green algæ. They would evidently replace the Crenothrix cocci in the culture tube as they did in the reservoir. These green algæ are the saviours of the water, for they oxygenate it and cause that organic food to disappear which Crenothrix utilises for its nutrition. The most successful artificial cultivations of Crenothrix yet obtained have been those which were instituted by Rullmann (1). They originated when the half-yearly purification of the reservoir which supplies the Bavarian town of Landshut revealed the existence at the bottom of the reservoir of a sediment of mud which demanded investigation. The mud owed its origin in the first place to the after-effects which followed a rapid growth of Crenothrix. Rullmann undertook an attempt to cultivate this organism artificially from the material supplied to him from Landshut. It was not found difficult to secure growth in many media. Thus good results were obtained by adding small quantities of ferric and manganese hydroxide to an infusion of hay. After sterilisation this mixture was inoculated with the reservoir water from Landshut. After three months a satisfactory growth of Crenothrix had taken place, but unfortunately certain moulds and other micro-organisms had also succeeded in establishing themselves in the water. Rullmann repeated Rössler's experiments by cultivating Crenothrix on small bits of sterilised bricks placed in high Petri-dishes, and again succeeded in obtaining an impure culture of Crenothrix. Successful results were also obtained when various organic constituents —e.g. small quantities of nutrient broth—were added to material containing Crenothrix. Further, no difficulty was experienced in securing an ample growth of Crenothrix on solid media, for it developed equally well when its nutrient medium was stiffened with gelatine or agar.

Having thus secured a plentiful growth, Rullmann made many efforts to obtain pure cultures, but in this he was not favoured by fortune although he tried all the methods known to be available for procuring the isolation of bacteria. Rullmann's conclusion was that the multiplication of Crenothrix polyspora is possible in artificial liquid and solid nutrient media, with or without the addition of organic material. We cannot, however, agree *in toto* with this conclusion, or at any rate we must explain where it is liable to error. The important point on which a verdict must be obtained is whether or not Crenothrix makes use of the organic material at its disposal. While a definite amount of organic material may not have been added to a medium, that does not imply the complete absence of organic matter from the medium from which Crenothrix drew its supply of nutriment. In

8

Rullmann's experiments there was no single instance in which organic matter in sufficient quantity to support bacterial life was absent. This is obvious in the cases in which the following nutrient media were used : infusion of hay, nutrient broth, mangan-peptone, mangan-phospholactium, and water containing pieces of turf. On inspection it is seen that the same applies to the media in which Landshut water or Moldau water was used. It must be borne in mind that it is not a rare experience to find that a water which by chemical analysis appears to be pure, yet contains enough organic matter to support several thousands of saprophytic bacteria to each cubic centimetre, and it appears more than probable that the waters used by Rullmann were of this nature. It is obvious from his description that his best results were obtained when organic matter was specifically added, and there is no proof that in his more moderate results organic matter was absent. In dealing with organisms of this class which are saprophytic, it is necessary to remember that they can subsist on an extraordinarily small quantity of organic matter, and before Rullmann is in a position to state that they can subsist without the presence of organic matter, growth of Crenothrix would have to be obtained in media from which every trace of organic matter had been carefully excluded. In the case of some of the other iron-bacteria it has been definitely established, as is shown below, that their growth is dependent on the presence of organic matter, and it seems to the

present writer that the results obtained by Rullmann
for Crenothrix polyspora strengthen rather than weaken
the conviction that the same is true also for this micro-
organism. With regard to the iron-salts added by
Rullmann to most of his cultures, there seems no
ground for the assumption—which this writer did not
make—that they were indispensable for the growth
of Crenothrix, although, of course, in common with
certain moulds that were present in the same cultures,
this representative of the iron-bacteria exercised an
attracting influence on ferruginous compounds that had
been added to the cultures.

ARTIFICIAL CULTURES OF CLADOTHRIX DICHOTOMA

It has already been noted that Cladothrix dichotoma
is a name for a group of closely related organisms,
some of which are quite indifferent to the presence of
iron in the water. It is not therefore surprising that
the most successful artificial cultures of the iron-bacteria should have been accomplished with this adapt-
able organism. Büsgen (1) was the first to succeed in
this direction. He found that Cladothrix grew readily
in liquid and solid media which contained flesh-extract
in very small quantity (about ½ gram to 1 litre water).
By solidifying with gelatine and making plate cultures
separate colonies were obtained which made possible
the initiation of pure cultures. A very complete and
accurate account of the growth of Cladothrix in
various liquid and solid media is given by Hoeflich
(1). I have confirmed the results obtained by Büsgen

and by Hoeflich. It is interesting to note that under artificial conditions the lines of development followed by the British varieties do not in any essential differ from those followed by the Continental varieties. In all liquid media growth is made perceptible by the appearance in the liquid of numerous grey specks which when present in the liquid in large numbers give it a somewhat turbid appearance. These little specks usually congregate at the surface, but when gently shaken they fall to the bottom and remain there, forming in time a light grey flocculent deposit. When put under the microscope each speck is seen to be made up of numerous filaments of Cladothrix (Fig. 45) radiating from a mucilaginous particle. It is not advisable to boil the culture to make it sterile, for Cladothrix is strongly aerobic, and the process of boiling rids the water of enough oxygen to prevent the growth of Cladothrix. If pure cultures are desired a dilute flesh-extract solution should be solidified by the addition of just as much gelatine and no more (about $4\frac{1}{2}$ per cent) as is necessary to cause the medium to set solid on cooling.

The following facts are taken from Hoeflich's thorough work on the pure artificial cultures of Cladothrix dichotoma (1) :—

1. *Broth-culture* ($\frac{1}{2}$ gram flesh extract to 1 litre water).—He records the following :—

(1) The addition of peptone, sodium chloride, etc., acts very deleteriously on the growth of Cladothrix.

(2) Small grey specks appear in twenty-four hours when cultivation is made at 30° C.

(3) After a few weeks, when left undisturbed, a soft downy layer is formed at the bottom of the culture vessel—usually an open beaker—while in the liquid small motile flecks are observed which also sooner or later fall to the bottom. He does not mention the cause of the motility of these little spots that roam about the fluid, but as each is made up of filaments of Cladothrix and each filament is made up of cells, we may conjecture that many of these cells possess cilia, the lashing of which causes the motility.

(4) Colonies made up of a conglomeration of the flecks are formed on the walls of the culture vessel.

(5) On the surface there may develop a rapid multiplication, resulting in the formation of a greyish, soft, downy layer about half a centimetre deep. (This layer is not made of coherent material and is chiefly distinguishable from the rest of the fluid by its difference in colour and its greater opaqueness.)

(6) Occasionally on the surface Zoogloea-formation takes place when the mucilaginous secretion is great. When this takes place a more or less firm coherent layer is formed at the surface. This layer is opaque, and brownish-yellow in colour.

(7) The organism is strongly aerobic, but independent of light.

2. *Gelatine-cultures* (45 grams gelatine to a litre of broth medium).—Reaction should be neutral, or very slightly acid.

(a) *Gelatine - plates.* — Colonies are very small, cloudy-whitish, of irregular form, and sunk in the

gelatine. They seldom cause the complete liquefaction of the gelatine, but in a week or two after
inoculation a partial liquefaction is noticeable.

(*b*) *Gelatine Stab-cultures.*—The gelatine is liquefied, and in the liquefied mass small whitish specks
appear. In addition copious growth took place in
the following media :—

3. *Flesh-water Peptone Broth.*

Flesh	.	.	500 grams
Peptone	.	.	10 ,,
Sodium chloride	.	5 ,,
Water	.	.	1 litre

4. *Flesh-water Peptone Gelatine.*

The same as 3 but solidified with 10 per cent
gelatine.

5. *Flesh-water Agar-agar.*

The same as 3 but solidified with $1\frac{1}{2}$ per cent
Agar-agar.

6. *Flesh-water Glycerine Agar-agar.*

The same as 5 but with the addition of 2-$4\frac{1}{2}$ per cent
glycerine.

7. *Liquid Blood Serum* of various animals.

8. In addition to the above media, Hoeflich found
suitable nutrient substances in *Milk, Potato, Potato-
water, Hay-infusion*, and *Straw-infusion.*

ARTIFICIAL CULTIVATION OF LEPTOTHRIX OCHRACEA

Completely successful results have also followed
the attempts at the artificial cultivation of this widely
distributed organism. Winogradsky (1) was the first

to suggest a method which was successful in his own experiments, but all attempts at the same method on the part of other investigators have been uniformly unsuccessful. This investigator put into a glass cylinder a handful of hay—previously cooked in a large quantity of water—and added to it some water containing some freshly precipitated ferric hydroxide. After 8 to 10 days the glass walls were completely covered with a coating of a rust-coloured deposit, and Zooglœa-masses of the same colour appeared on the surface. These phenomena were stated to be due to a number of organisms among which Leptothrix ochracea never failed to appear. It will be observed that ferric hydroxide was added under the supposition that the presence of this compound was an absolute necessity, but Winogradsky does not seem to have attempted to secure a growth of Leptothrix in the absence of the iron compound.

Next in order we find a method by Adler (1).

His culture fluid consisted of the tap-water of Prag, to which 0·05 per cent ferrous ammonium citrate had been added. In a few days after inoculation yellowish, flocculent masses of Leptothrix, together with Clado-thrix, the Flagellate Anthophysa, and certain In-fusoria appeared in the medium. In commenting on this method Molisch asserts that by following the same procedure with the *hard* water of Vienna, he was not able to procure successful results. In Prag the water is *soft*, and has previously flowed over a large tract of moorland country before arriving at its destination. It is thus well adapted for the growth of micro-organisms of the type of these under consideration. It will be noticed that the cultures both of Winogradsky and of Adler were not pure. The acquisition of pure artificial cultures was reserved for Molisch (2). He first added to the Prag drinking water—the raw water of the town apparently always contains Leptothrix ochracea in its flora—various organic salts in small quantities, in order to ascertain which favoured the growth of Leptothrix ochracea. He then ascertained that the addition of manganese-peptone[1] to this water resulted in the dominance of Leptothrix ochracea. Similar results were also obtained by the addition of ferrous ammonium citrate, or peptone, or manganese carbonate. Having, by the addition of one or other of these salts, obtained a good growth, Molisch inoculated some of the culture-fluid, now of course containing large numbers of Leptothrix, into a medium made up as follows :—

Water   .   .   .   1 litre

Manganese peptone   .   0·5 gram

Agar .   .   .   .   10 grams

He succeeded in obtaining colonies in this solid medium and thus was able to sub-culture the organism in a pure condition, uncontaminated by the presence of other micro-organisms.

In the attempts to obtain pure cultures, it is in-

---

[1] Manganese peptone is a pharmaceutical preparation, and contains manganese, peptone, sugar, and some alkalis. The amount of manganese is 4 per cent.

teresting to note that much trouble was encountered by Molisch in getting rid of a coccus, numbers of which were found adhering to the Leptothrix threads, thus preventing the latter from being present in a pure form in the cultures.   When, however, the culture medium contained 2 per cent peptone, Leptothrix developed *motile* threads, and these were free from the cocci.   We see here a state of motility developed in an organism which, so far as was known, had never hitherto developed motility when growing under natural conditions.   Further, if one of the colonies of Leptothrix on an agar-plate be gently pressed down with a cover slip, numerous rods are set free from the colony, which at once begin to exhibit motility.

The colonies in agar of this organism are rusty-brown, roundish bodies, the colour in this case being due to the deposit on and around them of the oxide of manganese that is present in the culture medium. In form and structure the colonies are similar to those formed normally by the majority of the lower bacteria on agar media.   The rods liberated from these colonies by pressure on them with a cover slip are 0.6μ-0.8μ in width, and 2μ-14μ long.   They move in the direction of their long axis, and also exhibit a rotary movement.   They are ærobic, for they do not sink deep into the surface of the agar. In gelatine stab-cultures which Molisch was successful in growing, specks appear in the gelatine which are brown in colour where the supply of oxygen is plentiful, but colourless where this supply is deficient. The gelatine is gradually liquefied, although the liquefaction proceeds very slowly. The temperature limits of growth are between 5° C. and 40° C., the optimum being 23° C.-25° C. The organism grows in diffuse daylight, in the dark, and in direct sunlight. If the culture-medium contains manganese, the threads of Leptothrix which grow in this medium appear covered with a store of manganese hydroxide on their membranes. The other interesting results of these artificial cultures have been described already in dealing with the life history of Leptothrix ochracea. These experiments of Molisch mark an important point in the history of the researches on the iron-bacteria. Campbell Brown (1) had already proclaimed the dependence of the iron-bacteria on an adequate supply of organic matter, and a complete vindication of the truth of this fact was supplied for the first time by Molisch's experiments on Leptothrix ochracea. We shall return to this subject in dealing with the physiology of the iron-bacteria.

ARTIFICIAL CULTURES OF SPIROPHYLLUM FERRUGINEUM

The writer has obtained a slight intensification of growth in the case of Spirophyllum ferrugineum by adding freshly precipitated ferric hydroxide to well-water, and, after inoculation with water containing Spirophyllum ferrugineum, exposing the culture at room temperature to diffuse daylight. After two or three weeks a flocculent red deposit, made up almost exclusively of Spirophyllum ferrugineum, was formed at the bottom of the culture-flask, and the bands were in many cases thickly covered with conidia. These experiments were not continued further. In view of later developments in the physiology of the iron-bacteria, it seems probable that the determining factor of growth was the organic matter contained in the well-water, and not the ferruginous material added to the culture. Although the bands had undoubtedly attracted iron to their surface, in view of what we now know of the iron-bacteria, which was not known when the experiments were made, this is to be explained by the fact that the iron-bacteria exercise a chemiotactic influence on iron compounds (see Chap. VIII).

Hitherto no successful attempts have been reported in regard to the artificial cultivation of *Gallionella ferruginea* and of *Clonothrix fusca*.

Review of the Results of the Artificial Cultivation of the Iron-Bacteria

In the case of two of the iron-bacteria, namely, Leptothrix ochracea and Cladothrix dichotoma, pure artificial cultures have been obtained, out of the investigation of which the extremely important truth has emerged that these organisms are completely dependent on the presence of organic matter for their sustenance. Further, for the cultivation of these two organisms, it was not found that a supply of ferruginous material was indispensably necessary. It is evident that our consideration of the physiology of the iron-bacteria must be based on the bed-rock of the investigations on pure artificial cultures. In regard to the impure cultures of Crenothrix which Rössler and Rullmann succeeded in establishing, and in regard to my own impure cultures of Spirophyllum ferrugineum, there was no instance of growth in which it could specifically be stated that no trace of organic matter was present in the culture medium, and in the case of Crenothrix there was abundant evidence that growth was very probably influenced by the presence of appropriate organic matter. Hence there does not seem any evidence to warrant the assertion that Crenothrix and Spirophyllum are in any way different, in regard to their physiological needs, to Leptothrix and Cladothrix. There can be little doubt that it is only the imperfection of our knowledge of the conditions which determine the growth of Crenothrix and Spirophyllum which prevents us from being as certain as to their physiological needs as we are in regard to Leptothrix and Cladothrix.

The iron-bacteria therefore fall into line with the vast majority of the bacteria that we are acquainted with. They derive their nutriment in the same way and do not stand in the same relation to ferruginous compounds that the sulphur bacteria do to the sulphur compounds. We shall see the important bearing of these results when we consider the physiology of the iron-bacteria in relation to the problems which the engineers have to encounter. It may, however, be stated here, that these facts show that problems associated with the combating of the excessive multiplication of the iron-bacteria are essentially similar to those which the biologist and engineer have to deal with when excessive multiplication of various algæ or protozoa has taken place in the waters under their supervision. In the case of the iron-bacteria, however, the problem is complicated by the fact that these organisms are endowed with a capacity for absorbing large quantities of organic matter to the molecules of which iron is attached, and with the possession of a sheath which is capable of retaining the iron that is liberated to the outside after the organisms have done with the organic matter to which it was originally attached.

CHAPTER VIII

THE PHYSIOLOGY OF THE IRON-BACTERIA

MUCH interest has been taken in the physiology of the iron-bacteria from the time that Cohn (1) launched the first theory as to their mode of life, up to the present day when it cannot be said that all the problems connected with their physiology are free from all obscurity. The interest is connected with the strange peculiarity which the organisms of this group possess of gathering upon their surfaces a large quantity of iron from the surrounding water. This iron is found deposited on the iron-bacteria in the form of the insoluble ferric hydroxide, the amount of this substance being often so great that the deposit is greater in volume than the organism itself. After death the sheaths or membranes laden with and penetrated by iron, sink to the bed of the stream in which they live, and gradually build up there thick layers of a rusty-red colour which have to be removed periodically as such layers constitute a source of obstruction to the flow of the water. A microscopic examination of the material composing these "ochre-beds" shows a collection of hollow tubes impregnated with the iron.

These are the remains of the iron-bacteria. It is evident that as other micro-organisms—or the vast majority of them—that are present in the same water do not collect iron salts with anything like the same avidity, there must be some physiological activity or some peculiarity of structure in the iron-bacteria which enables them to bring about such a state of matters. Cohn was the first to suggest an explanation. It is known that diatoms absorb silicon from the water in which they live to a far greater extent than do most other plants. They build up their enclosing shells with this substance. Cohn suggested that ferruginous salts were as necessary to the iron-bacteria as was silicon to the diatoms. He thought that the iron salts were *selected* by the iron-bacteria, that is, that the protoplasm of these organisms picked out ferruginous salts from the other constituents of the water and utilised them to strengthen their membranes. This explanation is plausible enough, but we know now that to carry on life this stiffening of the membrane is not only not necessary but is actually harmful because it entails a hindrance of growth and a cessation of the reproductive activity. It is now known that the organism carries on its activities *in spite* of the presence of the iron, and artificial cultivations are possible in media to which iron salts have not been added.[1] Cohn's suggested explanation was therefore

[1] Every organism requires a very minute amount of iron, and of course the iron-bacteria are no exceptions in this respect, and every natural water contains the small amount of iron needed.

soon discarded.   Next, Zopf (1, 2, 3) directed attention to the mucilaginous covering which invests
the true membrane of all bacteria and similar
organisms.   This covering owes its formation to a
mucinous change in the outer layer of the organism.
It forms an investment of a clinging nature around the
rest of the organism, and is not visible unless it has
been rendered opaque by special treatment.

Zopf held that iron compounds underwent oxidation
in this covering, the product of oxidation, namely,
ferric hydroxide, colouring the organism and giving
ultimately to it a far greater perceptible volume than
it would have if the covering remained transparent
and so invisible.   He described the process as being
analogous to that whereby jellies are artificially
coloured with various dyes.   This explanation is
fundamentally different to that of Cohn, in that metabolism plays no part in it, consequently the iron salts
do not enter inside the cell and enter into any reactions
with the protoplasm.

We had better defer a criticism of Zopf's hypothesis
until we have dealt with the various facts that have
been elicited within the last ten years.   It was soon
abandoned and gave way to the hypothesis which was
promulgated by Winogradsky, in 1888, and which is
found in every text-book of bacteriology which deals
with the iron-bacteria.

According to Winogradsky, the iron-bacteria obtain
their vital energy, that is the energy necessary for
growth, movement, multiplication, etc., not as most

organisms, including the majority of bacteria, do, by breaking down various organic compounds and utilising the energy thus liberated, but *by the oxidation of ferrous to ferric compounds.*

A short time previously Winogradsky had made a brilliant exposition of the part played by the sulphur-bacteria in securing energy by the oxidation of sulphides to sulphates, and, *by analogy*, had applied the same kind of explanation to the iron-bacteria. The reaction may be represented as follows :—

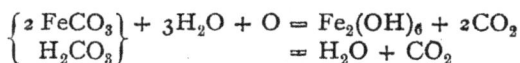

$$\left\{ \begin{array}{l} 2\,FeCO_3 \\ H_2CO_3 \end{array} \right\} + 3H_2O + O = Fe_2(OH)_6 + 2CO_2$$
$$= H_2O + CO_2$$

Here we have an exothermic reaction, that is, one the incidence of which is correlated with the liberation of energy. In nature solutions bearing iron in the soluble ferrous form spring up out of the bowels of the earth, and in such of these as come in contact with decomposing vegetable remains reduction is reckoned to take place with the consequent formation of ferrous bicarbonate, $FeH_2(CO_3)_2$. Now this substance is in solution, and, when oxidised, it changes according to the above reaction into the insoluble ferric hydroxide $Fe_2(OH)_6$.

It is asserted by Winogradsky that the iron-bacteria are responsible for the oxidation of the ferrous into ferric compounds, and that they thereby gain a certain amount of energy. He further stated that life was not possible to them if this source of energy were removed; that, in short, if ferruginous compounds are not present the iron-bacteria are not able to eke

9

out an existence.   If we examine this hypothesis we find that it does not bear close investigation.   In the first place, the acid carbonate $FeH_2(CO_3)_2$ is extremely unstable : after its artificial preparation it is found impossible to prevent its oxidation to $Fe_2(OH)_6$, which change takes place with great rapidity, and it seems difficult to believe that the iron-bacteria could snatch energy from such an unstable substance.   It is important to note that this oxidation takes place in nature whether the iron-bacteria are present or not. When we cite the additional fact that some of the most typical iron-bacteria have been grown in media devoid of iron the inadmissability of Winogradsky's hypothesis is evident.   His position is rendered still more untenable by the fact that the hypothesis was apparently advanced from analogy with the mode of life presented by the sulphur-bacteria, and no supporting facts were adduced that in any way strengthened his position.   For a long time Winogradsky's hypothesis was regarded as of the nature of a complete explanation in spite of the weakness of its props.   The first weakening blow was delivered by the publication of the results of Campbell Brown's (1) research, although this writer was concerned only in an incidental manner with the physiology of the iron-bacteria. During the course of a chemical investigation of the various ferruginous waters of this country, Campbell Brown instituted a comparison between ferruginous waters in which iron-bacteria appeared with those in which the latter were absent, and came to the conclusion that in no case were the bacteria present except in waters which contained an appreciable amount of iron in solution *in combination with organic matter of an acid character.* The acidity of such waters was invariably produced by the presence of carbonic acid. The growth of the iron-bacteria was not sensibly affected by the presence of bicarbonates. He came to the further conclusion that the occurrence of organic matter in solution does not necessarily imply the development of iron-bacteria, but in all cases in which development had taken place, this organic matter had iron attached to its molecule. Campbell Brown's summary is contained in the following words : " The slime organisms live on the carbon compounds in a soluble organic compound of iron, which penetrates into their substance, even into the inner tube. Whether the iron performs any function or not, the carbon compounds support life and the iron-oxide is necessarily deposited throughout the whole mass of living matter, but chiefly in the active parts." Here the fundamental importance of the presence of organic matter, and the subordinate rôle of the iron is emphasised for the first time. Campbell Brown was a chemist and was concerned only with the relationship of the " ferruginous slime " to the constitution of the waters in which this slime develops, and although he knew that the slime was caused by ferruginous organisms, these receive from him only a casual mention. It is to him, however, that we owe the first mention which was made of the

importance of the presence of organic matter in promoting the growth of iron-bacteria.

Later investigations have given abundant proofs
of the correctness of his theory. Six years after the
publication of Campbell Brown's work an important
addition to the subject was rendered by the publication of Molisch's (2) "Eisen-bakterien," for this
investigator succeeded in isolating Leptothrix ochracea
and sub-culturing it in artificial media. He was able
to show that Leptothrix will grow only so long as it
is supplied with organic matter, but that, failing this
supply, its cultivation is not possible. He proved
in the most conclusive manner that it was not
necessary to supply these organisms with ferruginous
salts. It is very strange that Molisch should have
made no mention of Campbell Brown's contribution,
although his knowledge of and access to his paper
is shown by his reference to it in connection with
another problem bearing on this subject-matter. On
the strength of the flood of light that was thrown on this
subject, a new theory was promulgated by Molisch.
In iron-waters the organic matter which is absorbed
and digested by the iron-bacteria contains an iron-
radicle. The organic matter is naturally absorbed in
solution and Molisch's theory is that the absorption
and digestion of the organic matter is followed by
the expulsion and precipitation of the iron which is
bound in the molecule of this organic matter. The
iron is caught in the sheath and there undergoes
oxidation with the resulting formation of ferric

hydroxide. The actual oxidation is thus made quite independent of the metabolism of the cell, the iron oxidising just as readily on the surface of the bacteria as it would on any other surface, such as a blade of grass or a stone.

When we examine this theory we find that it explains most at any rate of the facts bearing on the subject. Thus it is consistent with Campbell Brown's statement that iron-bacteria do not thrive except in waters containing an appreciable amount of iron in solution in combination with organic matter of an acid character. It is consistent with the finding of the ferric oxide not only in the sheath, but also in the cell membrane, and even inside the cell itself. Again, if we try to test the theory by the facts that were ascertained by Adler (1) we find that it emerges all the stronger after the test. This investigator found that the addition of camphor, alcohol, formic acid and other antiseptics to a culture containing Gallionella, in sufficient quantities to prevent that organism from multiplying, hindered considerably the rate of formation of ferric hydroxide in that culture. If, according to the theory, iron is thrown out of solution and spread out over the sheath and cell membranes, one can readily see that the oxidation of the iron would take place more rapidly than if the iron were still bound up in solution in the organic molecule. Still more is the theory consistent with the fact that not only is manganese present in the membranes of the iron-bacteria, but this substance may be found in greater quantity than the iron itself. Thus Jackson (1) found that in some specimens of Crenothrix there was present in the membranes 33·9 per cent manganese as against 14·4 per cent of ferric hydroxide. As pointed out by Molisch we have no instance in nature in which iron is essential to a plant or an animal and in which it can at the same time be replaced by manganese. The fact that the amount of iron, in proportion to manganese, which is found on the membranes varies considerably, and that this variation bears a certain amount of relationship to the respective amounts of these substances which are found in the water would make one suspect the essentiality of iron, quite apart from the absolutely conclusive evidence on this point which was supplied by the artificial culture of Leptothrix ochracea and Cladothrix dichotoma being found possible without the addition of ferruginous salts. Further proof of the replaceability of iron by manganese has been supplied by the researches of Beythien, Hempel and Kraft (1) and by those of Raumer (1). In fact, it would be quite as appropriate to designate these organisms manganese-bacteria as to name them iron-bacteria. It is the organic molecule in the water with which the protoplasmic molecule of the bacteria enters into combination, and such being the case it will naturally happen that in some cases the organic molecule will have iron in attachment, in others manganese. Hence the amount of iron or manganese on the membranes of the bacteria will depend on the constitution of the organic molecule with which the protoplasmic molecule has entered into combination.   If the organic molecule has no iron in combination the bacteria will thrive equally well as if the organic molecule were bound up with iron.   The deposit on their membranes will always contain quantities of iron or manganese in exact proportion to the amount of these substances in the organic molecules taken up by the bacteria.   Iron is thus not the coal which supplies the energy, but the slag which is left over when the whole process is finished, and the composition of the slag will vary in every stream in accordance with the variations in the organic content of the streams.   If we regard the iron as a by-product, the slag of our illustration, the results obtained by Raumer, Kraft, Beythien, and Hempel are readily understood, but if we regard it as a substance of vital importance, the coal of our illustration, the results obtained by these writers are not easy to understand.   So far all the facts support the theory that is based on the results obtained by Campbell Brown and by Molisch, and subsequent work has not produced evidence that can be regarded as affecting the groundwork of the theory.   It must, however, be stated that *all* the facts of the case are not covered by this theory.   In the first place, if the iron-bacteria are, *ex hypothesi*, saprophytic organisms like the majority of the denizens of ferruginous waters, it has to be explained why all these varieties are not equally coated with ferric hydroxide.   That some distinction

between the iron-bacteria and the other organisms
(protozoa, diatoms, etc.) exists is obvious at a glance
when ferruginous waters are microscopically examined.
The iron deposit on the sheaths and membranes of
the iron-bacteria cannot be wholly explained by the
retentive capacity of the sheaths, for the iron is seen
to have penetrated into the substance of the cells
themselves, and is found thickly laid over on organisms,
like Gallionella, which possess no sheath.   If we
assume Molisch's theory to be wholly true, it must
follow either that the iron-bacteria pick out these
organic molecules to which iron is bound, avoiding
organic molecules which contain no iron, or else that
the greater amount of iron on their surfaces in comparison with other organisms is due to the fact that
they take in more organic matter.   There is no
evidence in favour of the latter assumption, but
there is a probability that the first is correct, and if
so Molisch's theory is not complete, and must be
supplemented by the introduction of other factors.
We must, therefore, seek in other directions for the
explanation of this diversity in the amount of iron
collected by the different organisms in the same
ferruginous waters.   It must be remembered that
while the iron-bacteria are *facile princeps* in collecting
iron, they are by no means the only organisms on
which this metal collects in abundance.   Thus we
know from the researches of Klebs (1) that the
Zygnemaceæ (an order of green algæ) appropriate
to themselves from the surrounding water, aluminium,

iron, and chromium. Again, Gaidukov (1) found that the green algæ Conferva cladophthora, Œdogonium and certain Desmids collected iron in abundance when growing in ferruginous streams. The same investigator found this property to be characteristic of certain Protozoa, for example, Anthophysa vegetans. Lately Rullmann (1) in making artificial cultivations of the iron-bacteria, found in his cultures a certain unnamed thread fungus which was endowed with the same power of appropriating iron. Thus we have iron-collecting representatives from most diverse classes of organisms, and drawn from the animal as well as the vegetable kingdom. That the phenomenon is essentially of the same character in these organisms as in the iron-bacteria is rendered extremely probable by the detailed investigation which was carried out by Gaidukov (1) on a species of Conferva which inhabited a ferruginous organically contaminated pool of water. In this case the general appearance and changes in the water were the same as are seen in such waters when the iron-bacteria thrive in them. How are we to explain the pre-eminence of the iron-bacteria as collectors of iron? It is possible to see why the iron-bacteria should possess more iron than the green algæ or any of the chlorophyll-containing Protozoa. According to Binaghi (1) the oxidation of iron in iron piping is greatly expedited by the presence of carbonic acid in the water, and iron placed in water devoid of this gas does not change. If this be so, organisms devoid of colouring matter should be in a better position to collect iron than the coloured organisms, for they have no chlorophyll to eat up the carbonic acid given off during respiration, and as, according to Molisch, iron is being continuously thrust out of the cell there is better opportunity in such organisms for Binaghi's first reaction to take place.   This first reaction, according to Binaghi, is the formation of ferrous carbonate (see Chap. IX of this book).   In the case of organisms like the green algæ their carbon-assimilation has, during the daytime, the effect of removing the CO_2 from the immediate surface of their bodies, and the "local" supply of this gas must therefore be very limited.   Again as, according to most biologists, green algæ do not take in organic matters, and as the iron is bound to the organic matter, it is difficult to see how, if the theory applies to green algæ, the latter are in a position to store any iron at all.   And yet we know from the researches of Gaidukov (1) that some green algæ collect quite large quantities of iron on their bodies.   With regard to the colourless organisms other than the iron-bacteria, living in the same waters, it is very difficult to understand why all are not coated with iron to the same extent, for ex-hypothesi the absorption of organic matter is a part of the process of metabolism, and what applies to one kind of colourless organism applies to all organisms of the same class, for all alike must absorb organic matter to obtain sustenance, and in absorbing organic matter they must all absorb iron and consequently must more or less to the same extent subsequently throw this useless substance outside. But although some colourless organisms other than the iron-bacteria do collect iron, others do not, and therein lies the difficulty. It seems to me that we must here postulate the operation of another force, namely, that of a chemiotactic affinity of the iron-bacteria and the other organisms for iron compounds. The existence of this affinity on the part of iron-bacteria for iron compounds has not been investigated, but affinities of a similar nature are well known to occur in the life of most bacteria. Thus these organisms are attracted to corrosive sublimate even although the movement spells death to them. The proof of the existence of some such selection on the part of the iron-bacteria would solve the difficulties which were mentioned above. Under the influence of a chemiotactic affinity for iron, the iron-bacteria would absorb, principally at any rate, organic molecules containing iron in combination, and there would be absorption only in very small quantity of organic molecules devoid of iron in combination. We should understand then the diversity which exists in different colourless organisms, for those that did not possess any chemiotactic affinity for iron would not, except in a small degree, be tinged with the reddish-brown colour of ferric hydroxide. Lastly, with regard to colourless organisms it must be noted that if diversity exists—and it does—in the nature and extent of the mucilaginous covering which envelops the organisms, those containing the largest and the most adhesive coverings would hold the largest amount of ferric hydroxide. We know that the iron-bacteria are very well endowed in the way of sheaths or mucilaginous envelopes, and would therefore be well able to support a large quantity of ferric hydroxide. With regard to the green algæ some difficulties would still remain, but we cannot discuss them in this place. If we then sum up, we may state that the pre-eminence of the iron-bacteria in surrounding them-selves with ferric hydroxide is due to three facts, the existence of two of which is well established, and the third is highly probable.

1. They absorb organic matter for sustenance.

2. Probably they select organic matter with iron in combination owing to chemiotactic affinity for this substance.

3. They are surrounded by an excellent apparatus for detaining the ferric hydroxide which is formed on their surfaces, in the form of their mucilaginous sheaths or mucilaginous coverings.

The iron-bacteria are thus seen not to be an extraordinary group of organisms with a peculiar mode of existence, but are similar to many other colourless organisms that thrive in the same waters ; but that they have a special "talent" for collecting iron, due not to any outstanding qualities but rather to a combination of characters all possessed severally by various other organisms, all of which are favourable for the retention of iron. It is the combination of all three in strong degree which has made the iron-bacteria pre-eminent among the iron-collecting organisms.

CHAPTER IX

THE IRON-BACTERIA IN RELATION TO WATER RESERVOIRS AND THE CORROSION OF CONDUIT PIPES

A LTHOUGH there are still many outstanding problems in connection with the iron-bacteria, sufficient information has been gained with regard to their developmental phases and their physiological peculiarities to make the information at our disposal of value when applied to practical problems.

Slimy Streamers.—One of the visible signs of the development of iron-bacteria in conduit pipes is the formation in these pipes of slimy masses suspended in streamers from the walls of the pipes. Similar streamers also appear on the walls and the bottom—if the water is shallow—of the storage reservoirs. In these slimy masses ferric hydroxide collects and the engineer's choice of a remedy for its removal depends on whether this substance has been derived from the iron in the composition of the walls or from the iron in solution in the water. Inasmuch as the streamers form on protected as well as unprotected walls and on stonework and woodwork, it is evident that the water supplies at least a large percentage of the obstructing ferric hydroxide. The growth of these streamers inside conduit pipes results in a serious diminution in their carrying capacity if the streamers are allowed to grow unchecked. Now the amount of iron that is present in natural waters that are normally used for the supply of rural and urban districts is, as a rule, very small, as can be seen from the subjoined table which gives the amount of iron that was found in a few of the British sources of supply.

Oxides of iron expressed in parts per 100,000 :—

Water from Elan Valley	.	.	0·016.
,, ,, Welsh Valley	.	.	traces only.
,, ,, Ireland	.	.	0·004.
,, ,, Lancashire	.	.	traces only.
,, ,, Taff .	.	.	traces only.
,, ,, Montgomeryshire	.	traces only.
,, ,, Rivington .	.	.	0·016.
,, ,, Loch Katrine	.	.	0·002.
,, ,, Dee (Aberdeenshire)	.	none.
,, ,, Lake Vyrnwy	.	.	0·099.

In spite of this, however, the slimy streamers which form in the conduits and which grow on protected surfaces and on wood, stone, etc., must obviously derive the large amount of ferric hydroxide that collects on them and in their substance from the iron in solution in the water, for, growing on substances other than iron, they cannot obtain it in any other way. It must be borne in mind that although the amount of iron in the water is so very small—about one part in a million parts of water—and the amount taken in by the organisms causing the streamers very minute, still the process of absorbing iron and its oxidation is going on continuously, and, when oxidised, the slimy mucilage surrounding the iron-bacteria affords facilities for the storage of the insoluble ferric hydroxide, so that it is only a matter of time before the total amount of ferric hydroxide hanging on to the streamers is very considerable.

Another effect which follows the continued presence of the slimy streamers in the water is the gradual accumulation of organic matter which must take place after the death of the organisms covering the streamers. This follows from the fact that in the absence of chloro-phyll-containing organisms the dead bodies of the iron-bacteria and those of any other organisms that happen to have lived and died in the water tend to accumulate, and after various changes become the raw material of later generations of the same class of organisms. Hence the continued growth of the iron-bacteria must, in the absence of countervailing influences, tend to make a water steadily worse from the point of view of the engineer, owing to the steady increase which inevitably takes place in the organic content of the water. If algæ or other coloured organisms are also present, they exert a purifying effect on the water by means of the oxygen which they abundantly liberate. If, however, the conditions are unsuitable for the growth of coloured organisms, in course of time the percentage of organic matter will be so great that the saprophytic soil and water-bacteria are able to multiply rapidly and the water is no longer to be regarded as potable. Koch used to lay down the rule that the test of a very good drinking water was that it should not contain more than 100 bacteria per cubic centimetre. It is found in practice that a water in which these streamers are growing in abundance contains often several thousands of bacteria to the cubic centimetre, a sure sign that there is present far too much organic matter, otherwise the water could not have supported so many bacteria. When the streamers form in storage water the danger of a rise in the percentage of organic matter is always more or less present, but in the case of conduits, the flow of water is always sufficient to prevent the percentage rising to a dangerous level. The streamers and the black slimy lining in contact with the sides of tunnels, conduits, and culverts may hence be regarded as being injurious in three ways :—

1. They increase the amount of organic matter in solution in the water.

2. They decrease the bore of the conduit pipes.

3. They cause a decrease in the velocity of flow of the water.

In Campbell Brown's paper (1) analyses of the slime are given. In the various analyses the percentage of ferric oxide varies from 28 to 42 per cent, a considerable amount to take up from a water containing iron only to the extent of one in a million parts. And not only iron but also a large percentage of manganese in the form of manganese peroxide and manganese oxide is found in the slime, reaching in some cases to as much as 38 per cent.

When now the slime on the walls and the streamers in the water are microscopically examined, they are found to consist of huge masses of iron-bacteria. In the conduit pipes which supply Liverpool from Lake Vyrnwy the organism Gallionella (with probably Spirophyllum) preponderates, as is evidenced by Campbell Brown's figures, although he does not himself refer to any bacteria by name. Their mode of formation can be readily explained. The conidia of these organisms are present in abundance in ferruginous streams, and in growing become attached to the walls of the pipes. This first growing point becomes a focus of attraction, iron is absorbed, the mucilaginous sheath retains the oxidised iron, and so we have a thin thread thrust out into the stream, attached by one end to the wall. This grows in length and at the same time forms a mass of *conidia*, each of which, under favourable circumstances, is capable of developing into a new organism. Owing to the mucilaginous sheath which each organism forms, the daughter threads developed from the conidia do not separate, the whole mass remaining as a single large colony composed of a large number of organisms, each surrounded by its own mucilaginous sheath. To the naked eye this colony is visible as a single streamer of slimy material attached by one end to the wall, the other end waving free in the water. It can readily be seen how a streamer of this kind, by the

10

further multiplication of the individuals composing it,
and by the accretion of organic and inorganic particles
which would be delivered up to it by the flowing
water, would grow in volume until it would ultimately
interfere with the free passage of the water.

Tubercular Incrustations.—Another familiar phenomenon in the pipes is the formation of nodular
excrescences on the inner sides, where these are in
contact with the water.   While the causal relationship
of the iron-bacteria to the formation of slime is beyond
question, it is still a matter of controversy whether the
iron-bacteria exert any influence on the formation of
the tubercles.   These are limpet-shaped structures
that arise from the surface of the iron.   In time the
surface becomes studded with these "limpets," just
as a ship's bottom does with barnacles, and unless
cleared away they become confluent, the result being
a considerable decrease in the bore of the pipe and
the production of a softening effect on the iron, which
makes this material still more susceptible to the forces
of disintegration.   Each incrustation consists of a
cone-like structure and grows by the addition of concentric layers.   The central portion is black when
fresh and soft.   It often contains a little sulphide of
iron and becomes red on exposure to the air.   The
middle layers are often orange-red, and are composed
of ferric oxide.   Outside these layers are the outermost coverings arranged in concentric layers, and
composed for the main part of ferric oxide inter-
spersed with hard black layers of magnetic oxide of

iron. Conflicting views are held with regard to the mode of origin of the incrustations. In the first place, there is no general agreement that the iron of the incrustation comes entirely from the pipe, for some excellent authorities hold that the surrounding water also sends its contribution of iron. Again, while some hold that the mode of formation is purely chemical, others adduce facts to show that the iron-bacteria influence the formation. It must be conceded at once that the formation of tubercles may and does take place in the complete absence of the iron-bacteria or any other organisms. A simple experiment carried out by *Casagrandi* (1) proves that such is the case. This investigator passed water through U-tubes containing bits of cast-iron or steel, and found that if these were varnished, first of all blisters appeared after a time on the varnish. Then each blister ruptured and in its place a tubercle was formed. If the bits of iron or steel were not varnished, the formation of tubercles took place all the same, only this was not preceded by vesiculation. In this experiment it is evident that the formation of tubercles is a purely physical and chemical phenomenon, and the tubercle owes its origin to a change in the substance of the iron, not by accretion of iron from the surrounding water. The chemical changes involved in the formation of tubercles under these circumstances have been admirably worked out by Binaghi (1). We shall consider these changes first and ascertain whether they afford a complete explanation of the formation of tubercles *under all circumstances*. The iron in the piping is attacked by the CO_2 in the water, forming ferrous carbonate with evolution of H :—

$$Fe + H.OH + CO_2 = FeCO_3 + H_2 \qquad . \qquad i.$$

By the absorption of more CO_2 the ferrous carbonate is changed to the bicarbonate :—

$$FeCO_3 + H.OH + CO_2 = Fe(H.CO_3)_2 \qquad . \qquad ii.$$

Then the bicarbonate is changed to ferrous hydroxide :—

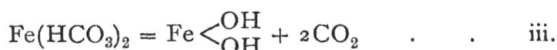

$$Fe(HCO_3)_2 = Fe{<}^{OH}_{OH} + 2CO_2 \qquad . \qquad . \qquad iii.$$

By oxidation the ferrous hydroxide becomes converted into ferric hydroxide :—

$$2Fe{<}^{OH}_{OH} + 2H_2O = 2Fe{\diagup}^{OH}_{-OH}_{\diagdown OH} + H_2 \qquad . \qquad iv.$$

Whether these reactions are absolutely or only partially correct there can be little doubt that all the constituents are present which are necessary for the completion of the oxidation of iron into the ferric hydroxide. Our main point, however, is to ascertain whether the process is affected by the growth of ferruginous organisms in the substance of the tubercles.

Casagrandi found that while some young tubercles contained iron-bacteria in their substance, these organisms were completely absent from others, and that, further, some tubercles contained ferruginous diatoms or other non-specific organisms. On the strength of his observations Casagrandi came to the conclusion that iron-bacteria do not exert any *apparent* influence on the formation of tubercles. While conceding the fact that the presence of ferruginous organisms is not a *sine qua non* for the formation of tubercles, it must, however, be stated that the probability of their aid in helping on the growth of the tubercle when they do happen to be present, is very great. One of the chief facts brought to light by Binaghi's investigation (1) was the dependence of the process on the presence of carbonic acid. Now any ferruginous organisms which may settle and develop on the tubercles will naturally furnish a constant supply of this acid, with the result that the change of iron into ferrous carbonate (No. i. reaction) will be expedited by their presence. Further, when, as a result of the changes outlined above, ferric hydroxide is formed, the mucilaginous envelopes of the iron-bacteria will prevent the escape of this substance, which will consequently begin to segregate round the tubercle.

The action of the CO_2 is one of solution, for the iron is actually eaten out of the pipe. This is seen by the fact that water traverses all the tubercular mass, and penetrates into the metal behind the mass. That the iron is actually removed is obvious from the fact that the tubercle contains only 70 per cent of iron. Hence one cannot agree with Casagrandi's conclusion, for it expresses only a partial truth. The part which the iron-bacteria play in the formation of tubercles is probably a very subordinate one, but such as it is, it expedites the change taking place inside the tubercle by secreting carbonic acid,

We may turn now to a third form of incrustation which derives its iron wholly from the surrounding water, thus differing essentially from the mode of collection peculiar to the tubercular masses. The rust layer of this type lies unequally thick on the surface, showing large and small swellings, and, in addition, finger-like extensions from the pipe into the surrounding water. When wet this layer is dark-brown, but when allowed to dry it assumes a rusty-red or yellowish colour. This layer is easily detached. *The walls themselves are not corroded.* In the case examined by Schorler (1)—pipes supplying Dresden—the layer appeared on a length of piping which had been coated with asphalt. On the removal of the asphalt the walls of the piping appeared quite intact. All the material for this layer must therefore have been supplied entirely from the store of iron in the surrounding water. In the course of examination a very interesting and important discovery was made by Schorler (1). The uppermost layer, which could be easily washed away, contained organisms identified by this investigator as Gallionella ferruginea. Underneath the uppermost layer was a firmer portion which contained no trace of Gallionella ferruginea, but there was in it an abundance of six-sided plates of a crystal nature. *Gallionella was found in the centre of some of these,* and Schorler came to the conclusion that their formation was due to a process of crystallization round Gallionella. We may therefore assume that the formation of these crystals takes place in much the same way as the nodules inside coal, which almost invariably contain at their centre a fragment of organic matter which has served as a rallying-point for the process of crystallization. Schorler expressed the opinion that the whole layer owed its formation to threads of Gallionella which settled on the walls of the piping where they attracted ferrous bicarbonate from the surrounding water, deriving energy from the oxidation of this substance to ferric hydroxide. While we cannot agree with this author in the *raison d'être* of the absorption of the bicarbonate, knowing, as we now do, that the organism derives its energy by the absorption and metabolism of organic matter, and that the oxidation is effected by the carbonic acid liberated during this process of metabolism, we can agree with him in this, that the organism was responsible for the accumulation of ferric hydroxide on the walls of the piping.

It is interesting to note that the same organism was found by Schorler under similar conditions in the conduit pipes which supply Bernberg, Essen, Frankfurt, and Teplitz. The results obtained by Casagrandi (1) in his examination of the Cagliari and other water supplies, appear at first to contradict the statement that the iron in the formation of these crusts and incrustations is derived from the water and not from the piping. Casagrandi states that there are hard waters which may contain ferruginous organisms, but there is no visible sign of their activity on

the pipes so long as the pipes are covered with a
calcareous deposit, but in the places where piping
coated with calcareous matter has been bared of this
deposit, ferruginous incrustations develop with rapiditv.
The inference drawn by Casagrandi was that the
ferruginous incrustation did not appear because the
iron in the piping had been covered by the calcareous
deposit, which seems to contradict Schorler's results.
He found an abundant growth of Gallionella on the
walls of pipes which had been coated with asphalt,
and a consequent abundant iron incrustation.   The
contradiction, however, is only apparent.   Schorler's
results show that ferric hydroxide will collect on the
walls of iron piping in spite of the protection offered
to the piping by a coating of asphalt, and that Gallionella is the active agent concerned in this act of
deposition.   Casagrandi shows that without carbonic
acid in the surrounding water, deposition does not
take place.   It is evident that the presence of the
calcareous deposit is an insuperable barrier to the
development of Gallionella, a barrier which is not
offered by asphalt.   We need not enter here into the
biological effects exerted on the iron-bacteria by
calcareous substances, but there is no doubt as to
their deleterious influence on the growth of these
organisms.   In the lime-lined walls, therefore, there
is a complete absence of any organic agent which can
draw the iron out of the water, and the pipe remains
clear.   Again, quite apart from the absence of the
iron-bacteria the presence of lime has the effect of

withdrawing carbonic acid from the water, and we have seen that iron pipes do not form incrustations in the absence of this gas. The whole body of evidence thus seems to favour the view that in protected pipes the ferruginous deposits there formed owe their origin to two factors, viz. the presence of iron in the water and the presence of ferruginous organisms which liberate a gas (CO_2) capable of transforming iron compounds into the soluble bicarbonate.

Spongy Disease of Iron.—Mention must be made of what is known as the spongy disease of cast iron. This is due to the attacks of acids derived from the water or from the soil, and is a process resulting in a slow solution of the iron with which the acids come into contact. In this operation the iron-bacteria have no part. In connection with this disease it is interesting to note that in essentials the gradual exhaustion of the iron is effected in the same manner as when iron-bacteria are responsible, namely, by the action of acids.

We may now conclude this section by summarising the results so far as they concern the iron-bacteria :—

1. *Slimy ferruginous streamers* in the water owe their existence entirely to the activities of iron-bacteria and other organisms of a similar nature.

2. *Ferruginous Tubercles.*—Iron is derived chiefly from the iron in the tubing, partly from the water as the result of activity of the iron-bacteria which effect local enrichments of CO_2 and thus facilitate the change of the iron into the soluble ferrous bicarbonate, $FeH(CO_3)_2$.

3. *Ferruginous Incrustations of a more or less Lumpy Nature.*—This kind lies unequally thick on the surface, is dark brown in colour, red on exposure, and beset with finger-like processes extending into the surrounding water.   In such cases the walls are not corroded.   The iron-bacteria, with possible help from other organisms, are responsible.

4. *Change of the Iron into a Mass of a Spongy Nature.*—The iron-bacteria play no part in producing this change.

CHAPTER X

THE TREATMENT OF WATER INFECTED BY IRON-BACTERIA

METHODS of Treatment.—It is obvious that the method of treatment will depend on the nature of the deposition of the incrustation. The engineer must first of all determine to which of the four classes of incrustations his own particular case belongs. We can here do no more than indicate the principles which should be relied upon to determine the treatment of particular cases.

1. *Spongy Disease of Iron.*—If the deposit on the iron is of a spongy nature (class 4 above), it is clear that there can in most cases be no remedy except the drastic one of taking away the pipes altogether and replacing them by new ones. Owing to its porous nature the iron has been dissolved out because acid-containing water has soaked in. In tubes affected by this disease the amount of iron may sink down to 50 per cent and under. To render the iron immune from further attack, either it must be made less porous by being coated in such a way that the entrance of water is not possible, or else the destructive agents, namely, the acids in the soil or water, must be removed altogether. The latter remedy is, of course,

not possible on a large scale, but if a perfectly non-porous coating can be laid on the piping, the softening process will cease. If, however, the coating has any blemish, the arrest of the softening is merely temporary, for if there is a spot at which water can get in it travels from the point of entry in all directions, and the softening process proceeds anew over all the parts.

2. *Removal of Slimy Streamers.*—With regard to the slimy ferruginous streamers (class 1 above), the question of their removal must be governed by our knowledge of the physiology of the iron-bacteria. It is evident that the complete removal of these organisms is impracticable, for the finest gauze would not prevent the passage of reproductive cells that are so small that 100 of them drawn up in one row could easily march through the eye of an ordinary needle. Attempts must therefore be made to deprive the organisms of their food or else prevent their using the organic matter by changing the composition of the latter in such a way that they are useless to the iron-bacteria. For example, the oxidation of the organic matter renders it unsuitable as food. The organic matter which to some extent or other contaminates all natural waters is removed by the following agencies :—

1. Aeration of the water.

2. Growth of green plants, because oxygen is liberated by such plants on exposure to light.

3. The nitrifying bacteria,

4. Animals which feed on the organic matter on the surface.

And the same methods, so far as they are applicable, must be employed in the treatment of any large body of water, which treatment has as its aim the removal of organic matter. With the removal of organic matter the formation of slimy streamers gradually ceases to increase, and in time they disappear altogether. Of these agencies the aeration of the water is undoubtedly the one which is the most practicable, and, where the conditions allow of it, should be adopted. With regard to the second agency, the growth of green plants, it is necessary to point out that while these plants supply an abundance of oxygen into the water during their lifetime, their bodies after death add to the organic matter, so that caution must be exercised in growing them, lest their dead bodies accentuate the evil which the living plants were intended to remedy. The action of the nitrifying bacteria must be here explained. They exist in shallow waters and near the surface of deep ones. They get rid of organic matter by causing its oxidation into compounds from which the iron-bacteria and similar organisms can obtain no sustenance. Hence the development of the nitrifying bacteria must be encouraged. When water is filtered or subjected to aeration in any other way, these bacteria grow to enormous numbers on the filters. The treatment of raw water by filtration is therefore to be recommended for the elimination of the iron-bacteria. When

sewage is subjected to filtration the main object in view
is the destruction of the organic matter, for it is only
in this way that the microbes can be kept down.
The precautions with regard to the cleansing of the
filters are the same when the aim is the destruction of
the iron-bacteria as when the aim is the destruction of
sewage. The skin which forms on the filters must be
removed at periodic intervals, otherwise the conditions
become too "rich" for the well-being of the nitrifying
bacteria, and they become replaced by bacteria that
work more harm than good. It is not necessary to
inoculate the filters with the nitrifying bacteria, for
they are present almost everywhere, and if the conditions are good they will multiply fast enough.

In regard to the fourth agency, viz. the cleansing
operation of animals, particularly birds, which clear
the water of large solid organic detritus, decision must
be made as to whether these animals sweep away
more objectionable matter than they create by their
own filth. There can be little doubt that slimy
streamers cannot subsist in a water treated as described above, for the organic content of a water of
this kind will be so altered as to have become unaccept-
able to the iron-bacteria. This treatment may be
supplemented by another which will also introduce a
condition disastrous to the well-being of these organisms. It has been stated that the iron-bacteria do not
thrive in water in which all trace of acidity has been removed by the introduction of *lime* or of *soda*. We may
recall Campbell Brown's statement that satisfactory

results were obtained experimentally by the introduction of one or other of these two substances.
Also it may be recalled that Casagrandi produced the
same immunity by lining the inside of the pipes,
through which the water passed, with a calcareous
deposit. The exact mode of treatment for the removal of slimy streamers must obviously vary according to the particular circumstances under which they
arise, but in all cases treatment should consist of
methods which have the two following ends in view :—

A. The removal of organic matter from the water.

B. The rendering alkaline either of the water or of
the parts on which the iron-bacteria grow.

The proportions, etc., of the substances that should
be employed in the treatment of the water to secure
the disappearance of the organisms without overdoing
the treatment must obviously be carefully worked out
before the application of treatment. In this book
we are concerned only with the principles which
should regulate the control of any water which has
shown signs of having developed the objectionable
streamers in undue numbers. Unfortunately, the
organism that is mainly responsible for the development of streamers is Gallionella ferruginea, which has
not hitherto been cultivated in an artificial medium,
and so it is not possible to state specifically the main
conditions of its growth. We can, however, feel
reasonably certain that these conditions differ in no
essential respects from those which must be imposed in
order to secure successful cultivations of Leptothrix

ochracea. We may therefore regard it as reasonably certain that Gallionella will not grow under conditions which we know to be unsuitable for Leptothrix. The nature of these conditions has been sufficiently indicated in the above description.

Tubercular Incrustations.—The experiments which have been carried out by Casagrandi have given ample proof that the iron-bacteria play such a subordinate rôle in the formation of tubercular incrustations that it is not worth while to take any special steps towards eradicating these organisms.

Iron Incrustations on Non-Ferruginous Surfaces.— As these incrustations are, according to Schorler, entirely due to the abstraction by bacteria and other organisms of iron from the surrounding water, it is obvious that the problem of their removal must be attacked in precisely the same way as when dealing with the removal of slimy streamers. As in the case of these streamers Gallionella is predominant, all that has been said above with regard to the removal of Gallionella from waters in which it formed slimy streamers, applies with equal force to the question of the removal of iron incrustations which appear on non-ferruginous surfaces.

Sudden Visitations of Crenothrix Polyspora and Cladothrix Dichotoma.—The experience of past years has been that the *sudden* visitations that occasionally occur in water reservoirs have been due either wholly to Crenothrix polyspora or to this organism in association with Cladothrix dichotoma, the latter playing, in the trouble, a somewhat subordinate rôle. We have a large body of knowledge concerning the conditions which determine the growth of Cladothrix dichotoma, and it may be stated that we shall not be far wrong in assuming that the same hold for Crenothrix. As these are similar to the conditions that hold for Leptothrix, we may assume that the conditions of growth that have been determined with precision for Cladothrix and Leptothrix hold true for Crenothrix and the remaining iron-bacteria. Knowing these facts it is possible to speak with confidence concerning the various ways that are open to us of dealing with the calamity :—

1. The complete removal of the organism would, of course, remove the trouble, but it is quite out of the question. The removal of the reproductive cells and the vegetative cells of Crenothrix and Cladothrix from a reservoir would not be possible without employing means that would practically annihilate all organic life in the water. The sterilisation of the water as a practical measure is impossible, and even if it could be accomplished the good effects would not be permanent, for the reproductive cells of various organisms, and among them those of the iron-bacteria, would once more penetrate into the water from all sides. There is no known specific which singles out the reproductive cells of the iron-bacteria for destruction and is relatively harmless to similar bodies belonging to other organisms. This method must therefore be eliminated as a practical measure.

2. Another method of attacking the problem is by ascertaining whether it is possible, not so much to destroy the reproductive cells as to prevent their germination. As has been repeatedly stated already, the two facts in connection with their growth which, from a practical standpoint, are of extreme importance are, first, their need of organic matter, and secondly, their abhorrence of an alkaline medium. If the methods delineated above for the removal of slimy streamers are applicable to the body of water under consideration, they should be employed. The engineer must set himself to the destruction of organic matter and the removal of all traces of acidity. The principles which underlie the methods by which these desirable results are brought about have been mentioned above in dealing with the treatment applicable to the removal of the streamers.

3. A third method of treatment consists in allowing the organism full scope for the short time when its growth is in full swing. This is on the principle of giving it enough rope to hang itself with. It is not perhaps sufficiently realised that when the multiplication of Crenothrix is in full swing during a "visitation," it is performing during this period an operation that should be the duty of the water engineer during normal times, namely, the destruction of the organic contamination which supplies the Crenothrix with the "sinews of war". So far as can be ascertained Crenothrix itself does not liberate toxic products that are a danger to the public health when

present in drinking-water.   In the case of Cheltenham, which is the severest case of a Crenothrix calamity recorded in this country, the water was not particularly harmful in spite of the extraordinary growth of Crenothrix which had taken place.   The growth of Crenothrix was incidentally instrumental in transforming a large amount of potentially dangerous organic matter into innocuous compounds.   This was the raw material for the growth of Crenothrix.   Its place was taken by a certain amount of debris due to the dead bodies of the defunct Crenothrix individuals. It is evident in the case of Cheltenham that the change wrought in the constitution of the water by the development of Crenothrix was not altogether of a harmful character, for, with the cessation of its active growth, which took place after a few weeks, the water assumed a refreshing appearance of greenness, which, of course, must have been due to the development in very large numbers of chlorophyll-containing organisms.   These latter organisms liberate oxygen into the water, and use up carbonic acid, and purify the water by their growth.   It can safely be asserted that in the case of Cheltenham, the water in the reservoir was in a far purer condition in respect to the needs of man after Crenothrix had ceased its active growth than at the time when this organism began its rapid multiplication.   The excessive growth of an organism that does not produce toxic products and yet uses up organic matter cannot be regarded as an unmixed evil, even although its temporary effects are somewhat disquieting.

4. As a fourth alternative, or supplementary treatment, must be considered the possibility of getting rid of that portion of the supply which is known to contain the organic matter which supplies the food. It has been shown in the previous pages that a calamity growth of Crenothrix has been associated with a supply, the water of which previous to storage has flowed over a moorland or a boggy country from which it has received that particular form of organic matter which was especially suitable for the needs of the iron-bacteria. If, therefore, the water supply of any urban district is drawn from different sources it becomes expedient to inquire whether the water from these sources may not possess different values in respect of their organic content. The determination of these values is a matter for a bacteriological, rather than for a chemical examination. The latter method is too gross to admit of that fineness of discrimination which is afforded by the analysis of the number and the character of the organisms which exist in the water. If the sources of supply are not all of the same character, it is well to know which are most likely to cause further trouble in the future. If it be found that such differences exist, and the water engineer draws his water from several sources, it would become incumbent upon him, at certain periods of the year, particularly during early summer, to cut off his reservoir from receiving its supply from the suspected area.

CHAPTER XI

THE IRON-BACTERIA IN RELATION TO THE FORMATION OF GEOLOGICAL STRATA

AN interesting scientific problem arises in the consideration of the question whether the iron-bacteria have played any part in the formation of geological strata, similar, for example, to the part played by foraminifera in the formation of limestone, or by vegetable remains in the formation of coal. On à priori grounds we should expect to find fossil remains of the iron-bacteria, for although their membranes are not composed of hard material, there is, judging by the appearance of the beds of iron-waters, a good deal of resistance to decomposition on the part of these membranes when they have been saturated throughout with ferric hydroxide. It is evident by the large deposits of the membranes of dead iron-bacteria in waters in which active multiplication of the iron-bacteria is very small, that their accumulation has been very gradual, and that years have elapsed in the case of many of them since these membranes enclosed living matter. That their preservation is possible over many years is a fact which cannot be called into question, but that the extension

in time can be stretched so as to penetrate into remote geological epochs is, on à priori grounds a more
doubtful matter.

We find the first mention of iron-bacteria in connection with rock formation in the writings of Ehrenberg
(1).   At the bottom of peat-mosses there is sometimes
found a cake of oxide of iron.   This is the familiar
bog iron-ore.   Ehrenberg had observed in the marshes
near Berlin a substance of a deep ochre-yellow passing
into red, which, on microscopical examination proved
to be composed almost entirely of Gallionella ferruginea.   He regarded this organism as a diatom,
and both Ehrenberg and Lyell regarded bog iron-ore
as composed of an aggregate of millions of these
organic bodies invisible to the naked eye.   The
organic bodies here referred to were the remains of
organisms which compose the ochre-beds of ferruginous
streams, which remains, are not, of course, in a fossilised condition.   That bog iron-ore should have been
formed from such beds is not remarkable, neither
would be the fact—if true—that the Gallionella membranes were found in bog iron-ore, for not only does
the ferric oxide lend a resistance to the membranes,
but also it is a well-attested fact that even organisms
of delicate texture, buried in peat, may resist decay
for untold generations.   Ehrenberg, however, did not
do more than *surmise* the agency of Gallionella and
its allies, in the formation of ferruginous rocks, and
the same may be said of the contribution to this
subject made by Winogradsky.   The next gleam of

light on the subject came from Molisch who carefully examined thirty-four iron-ores which he had obtained from various continental museums. He found the remains of iron-bacteria in three of these. Of these specimens two were derived from Siberia, while the third was of Prussian origin. The results of the examination are given by Molisch in some detail.

1. *A Bog Iron from Siberia.*—This broke easily into a fine ochre-coloured powder, and, while some parts of the powder were composed entirely of empty sheaths of the iron-bacteria, other portions were quite free from them. Further, all grades of condition were found between those two extremes.

2. *A Bog Iron-ore from Siberia.*—This ore was covered in parts by an ochre-brown thin layer which was made up entirely of the empty sheaths of iron-bacteria. The inner mass of the ore was completely free of organisms.

3. *A Bog Iron from Prussia.*—A few iron-bacteria were found and along with them were the remains of certain algæ belonging to the Chroococcaceæ or Palmellaceæ.

There can, therefore, be no question that many of the bog iron-ores owe their formation largely to the activities of the iron-bacteria. Molisch's results are interesting because they confirm what could be conjectured to be the case from an examination of the ochre-beds of present-day ferruginous waters. The same investigator extended his examination of ferruginous rocks, and in 1910 (Molisch (2)) gave the results of the examination of twenty-seven more specimens of iron-stones. He obtained a positive result only from a limonite of the class known as Raseneisenstein. Concerning this Molisch writes that the greater part showed no traces of iron-bacteria, but that here and there, inside holes, ochre masses were found, made up almost entirely of fragments of the sheaths of Leptothrix ochracea. Molisch's investigations have proved that in certain cases the threads of Gallionella ferruginea and of Leptothrix ochracea may persist for long periods. As the iron-bacteria are often active agents in the formation of the ochre beds of ferruginous waters, and as bog ore deposits are formed from such beds, these organisms must be regarded as rock builders. A more interesting point, however, is the question whether the iron-bacteria have contributed to the formation of older rocks than the comparatively modern bog-iron rocks. What, for example, has become of the bog iron-ores of the periods preceding the present? There can be little doubt that such rocks were formed and that their formation was expedited by the activities of the iron-bacteria. Mr. Tyrrell, of Glasgow University, is of the opinion that the Clayband Ironstone is the bog iron-ore of the Carboniferous period. If so it is not improbable that remains of the iron-bacteria would be found on careful examination of such rocks. Hitherto, however, no structures recognisable as the fossilised remains of iron-bacteria have been discovered in rocks of older date than bog iron-ores. In connection with these ores it must be borne in mind that modern bog ore sediments do not always contain iron-bacteria, for the reason that the oxidation of ferruginous compounds is, in many cases, accomplished by purely physical and chemical agencies in which the iron-bacteria play no part. Further, as Schorler has conclusively demonstrated, it is possible for the iron-bacteria, which have been largely instrumental in the formation of modern incrustations inside iron pipes, to be altogether destroyed very shortly after the completion of their activities. As, therefore, there are cases among modern deposits and incrustations of ferric hydroxide in which there is no trace of the organisms which have been partly instrumental in their formation, so we may conjecture that the evidence of their organic origin will have been destroyed in a still larger number of cases among the rocks of to-day which have been evolved from the bog ore sediments of former geological epochs.

We may turn now to another line of attack in which it is sought to prove that iron organisms have, by their activities, influenced the percentage of iron in the constitution of various ferruginous rocks. The currently accepted opinion with regard to the origin of iron in iron-ores is that this metal was introduced by the injection into various strata, under huge dynamic influences, of molten ferruginous matter. In the case of the majority of the iron-ores there can be little doubt that the iron in them has been so derived. It is urged, however, that cases are in evidence in which the probability is very great that a portion of the contained iron has been derived from other sources. The writer has made a study of ferruginous rocks in which organic fragments were found plentifully distributed, in order to ascertain whether evidence would be forthcoming which would shed some light on this question. It seemed advisable to approach the matter from a biological standpoint in order to supplement the evidence adduced from researches on the geological side. To this end the organic fragments in which certain ferruginous ores are plentifully distributed were submitted to a microscopical examination in order to ascertain whether they contained any micro-organisms. If they did it was considered probable that these would show indications of the nature of the medium in which they lived. Through the kindness of Mr. W. Thorneycroft I have been able to examine a large number of thin slides of iron-ore from various parts of the country where the ore is worked on a commercial scale, and have subjected these to examination. It is not to be expected that all the organic fragments that are seen in the ores were, when engulfed, free from the ravages of the micro-organisms that attack organic matter. Some must have been in a state of decomposition when engulfed, and decomposition implies an organic agent of decomposition. Further, it was not unreasonable to expect to see traces of the micro-organism which was the agent of decomposition. Among the rocks under observation a successful result followed the examina-

tion of the Frodingham Ironstone, which is a ferruginous, fossiliferous, limestone with distinct oolitic
structure.

The bulk of this ore is a hydrated peroxide or
limonite, and, according to Kendall (1), not very rich
in iron. The ore is distinctly bedded, and this is
decidedly increased by the occurrence of calcareous
bands which are of a much lighter colour. These
bands, which graduate on both sides into the ore,
contain numerous fossils, and so does the ore itself,
the same species being found in both. A search in
these fossils revealed the presence, not of iron-bacteria,
but of an iron-secreting thread fungus which has been
described elsewhere in detail (Ellis (9)).

The point to note in the present connection is the
fact that the mould, which has been engaged in the
decomposition of the numerous organic fragments of
this ore, showed a deposit of iron on its membrane,
and it is possible to come to a decision on biological
considerations as to whether or not this deposition
was laid down during the lifetime of the organism.
The reason for this lies in the fact that the deposition
on the membranes of modern iron-bacteria shows
certain characteristics, the nature of which is determined by the age of the organism and the length of
time that has elapsed since the deposition was laid
down. In the fossil mould *the deposit of iron on its
membranes showed the same variation in the nature of
the deposit.* This indicates that this fossil mould had
collected the iron on its membrane in precisely the

same way as do the iron-bacteria to-day. That is to
say, it must have lived in a ferruginous medium, and
the iron in it was collected during its lifetime.

If the iron on the membrane of this ferruginous
mould had been placed there by subsequent infiltra-
tion, the deposition would have been of a uniform
character on all the threads of the mould. We are
therefore driven to the conclusion that the waters in
which throve the organisms whose fossilised remains
are found in the ironstone were highly ferruginous
—although not so highly ferruginous as to make
organic life impossible—and that the first stage in the
process of fossilisation had resulted in the formation
of a bed that was distinctly ferruginous in character.
There can be little doubt that the bulk of the iron in
this rock was formed as the result of the injections of
molten metal that have subsequently been forced into
it. The point that is emphasised, however, is that the
rock was already ferruginous when this subsequent injection took place, for without this assumption it is
not possible to account for the facts of the case. We
may conjecture from the state of the mould the nature
of the conditions which prevailed at the spot where
the Frodingham Ironstone was laid down. The
water was shallow, boggy, and of a highly ferruginous and organic nature. Although ferruginous, this
water was not so highly charged with iron as to inhibit
the growth and multiplication of other organisms
which had no attraction for iron. After their death
the remains of the organisms which lived in this water

sank to the bottom of the shallow water, and the same bed would also be the resting-place for the remains of the larger land animals that lived in bogs of this character. The organic matter in the bed would undergo decomposition, and amongst the agents of decomposition is to be included a ferruginous mould, on the membranes of which is found a coating of iron of the same character as that found in modern iron-bacteria.

Sorley (1) advances as an argument in favour of his general theory—which is not here in dispute—that numerous living organisms could not have existed in waters charged with iron to such a degree as is found in the iron-ores of the Secondary Rocks which contain fossils. This argument, if true, would of course militate against the view that any organic influences had been brought to bear on the formation of any iron-ore. But the argument is obviously untenable because it is not the water in which the organisms thrive which is fossilised, but the bed on which the water rests. The water in which the iron-bacteria live contains, roughly, one part of iron to a million parts of water, but the bed underneath may contain iron to the extent of 60 or 70 per cent. It is this bed which becomes bog iron-ore, and not the water above it. As explained, the organisms first live in the water, and then after death sink to the bottom.

The same problem has been attacked from still another angle by Herdsman (1). This investigator has sought to prove the organic origin of the sedimentary

ores of iron by drawing attention to the phosphoric
nature of all these rocks.   While the phosphorus in
ore is of organic origin, this does not necessarily indicate
an organic origin for the ore itself.   It does, however,
constitute a sufficient basis for the prosecution of an
investigation as to the extent to which organisms have
contributed to the formation of ferruginous rocks.
The following observation by Herdsman is worthy
of note because of its bearing on the subject.   An
iron-ore placed on a shelf in a coach-house became
coated with the organism Merulius lacrymans—
familiarly known as "dry-rot".   This assumed a red
hue and threw off spores.   I can vouch for the
identity of this species and the red hue mentioned by
the author, as the specimen was sent to me by Mr.
Herdsman for identification.   The assay of the spores,
according to this writer, possessed the same iron
contents as the ore itself.   He does not say whether
the quality or the quantity was the same, but it is
biologically a remarkable phenomenon that such in-
tense absorption, out-rivalling the iron-bacteria, of
ferric oxide should have taken place on the part of
this organism.   As chemiotactic irritability for certain
specific substances is a common phenomenon among
the lower organisms, there is no reason to doubt the
substantial accuracy of Herdsman's observation.   This
writer's position with regard to the origin of the iron-
ores of Swedish Lapland is not apparently accepted
by the majority of other investigators, but there can
be little doubt the line of investigation traversed by
him is one which deserves careful consideration.

We may now summarise the points in favour of organisms having influenced the formation of iron-ore by their activities during life, and by the accumulation of their bodies after death.

1. Molisch has found four bog iron-ores which, either in parts of, or on the whole of, the surface, or, in the case of one ore, throughout its substance, contained abundant remains of Gallionella ferruginea and Leptothrix ochracea, two of the best known of the iron-bacteria.

2. The formation of iron-incrustations on iron-surfaces exposed to water is indubitably influenced by the activities of the iron-bacteria.

3. The formation of ferruginous sediments in boggy places is indubitably partly due to the same organisms, and the difference between such a sediment and bog iron-ore is one of degree and not of kind.

4. Schorler has shown that in certain cases in the formation of iron-incrustations, the causative organism, namely, Gallionella ferruginea, is present in abundance in the form of its dead sheaths on the surface layers, but that these sheaths disappear entirely a little below the surface.

5. My investigations have shown that one micro-organism found inside organic fragments from the Frodingham Ironstone possessed iron on its membranes, so disposed that it was possible to conclude that this iron was laid down during the lifetime of the organism. This proves that the waters, the deposited material of which laid the foundations of the Frodingham Ironstone, were highly ferruginous.

6. Herdsman calls attention to the greedy activity with which Merulius lacrymans absorbs iron.

7. Many iron-ores are highly phosphoric. This phosphorus must of necessity be organic in origin, and the possibility that a certain portion of it was derived from the remains of organisms which possessed the same physiological attributes as the iron-bacteria is not altogether excluded.

8. At the present day the iron-bacteria are responsible for the formation of ferruginous sediments which probably form themselves by slow degrees into iron-ores. Our present knowledge is still somewhat meagre, and it is not possible to come to definite conclusions as to the exact extent to which organisms have taken part in the building up of ferruginous ores. Enough has been shown, however, to make us beware of assuming that in no case have organisms played any part in the foundation of ferruginous ores.

INDEX

PRINTED IN GREAT BRITAIN AT THE UNIVERSITY PRESS, ABERDEEN